T0135872

Analysis of
geometrically non-linear models
for contact with dry friction

D i s s e r t a t i o n

zur Erlangung des akademischen Grades
doctor rerum naturalium
(Dr. rer. nat.)
im Fach Mathematik

eingereicht an der
Mathematisch-Naturwissenschaftlichen Fakultät II
der Humboldt-Universität zu Berlin

von
Diplom-Mathematiker Florian Schmid
geboren am 20.11.1975 in Oberndorf am Neckar.

Präsident der Humboldt-Universität zu Berlin
Prof. Dr. Dr. h.c. Christoph Markschies

Dekan der Mathematisch-Naturwissenschaftlichen Fakultät II
Prof. Dr. Wolfgang Coy

Gutachter:

1. Prof. Dr. Alexander Mielke

2. Prof. Dr. João A.C. Martins

3. Prof. Dr. Christof Eck

eingereicht am: 30. April 2008
Tag der mündlichen Prüfung: 2. Juli 2008

Bibliografische Information der Deutschen Nationalbibliothek

Die Deutsche Nationalbibliothek verzeichnet diese Publikation in der
Deutschen Nationalbibliografie; detaillierte bibliografische Daten sind
im Internet über http://dnb.d-nb.de abrufbar.

ISBN 978-3-8325-2005-2

Logos Verlag Berlin GmbH
Comeniushof, Gubener Str. 47,
10243 Berlin
Tel.: +49 030 42 85 10 90
Fax: +49 030 42 85 10 92
INTERNET: http://www.logos-verlag.de

Preface

Diese Arbeit entstand in drei Städten: Stuttgart, Lissabon und Berlin. Mein erster Dank geht an Prof. Dr. Alexander Mielke für die vertrauensvolle Zusammenarbeit über all die Jahre, die Weitsicht in der Betreuung und die Unterstützung bei der Umsetzung des Aufenthaltes in Lissabon. Alexander Mielke nahm sich immer Zeit und liess mir genügend Freiräume, die eigene Arbeit voranzubringen und zu gestalten.

Mein nächster Dank geht an Prof. João Martins der mir in Lissabon eine ganz andere Sicht auf die eigene Arbeit gab.

Unterstüzt wurde ich in meiner Forschungstätigkeit durch Projekte der DFG und dem Förderprojekt "Smart Systems" der Europäischen Union.

Korrektur gelesen haben Vanessa, Dorothee und Carsten und sie wissen, dass ich ihnen dafür dankbar bin.

Dankbar bin ich auch für die vielen Menschen, denen ich in diesen drei Städten begegnet bin und von denen ich mindestens genausoviel gelernt habe, wie durch meine Arbeit selbst.

Ein letzter aber vielleicht der wichtigste Dank geht an meine Eltern und meinen Grossen. Ohne ihre Unterstützung würde ich wohl nicht bis zu diesem Punkt gelangt sein.

Berlin, den 30. April 2008   Florian Schmid

Contents

Preface   v

Contents   vii

1 Introduction   **1**
  1.1   Modeling   2
  1.2   Assumptions and main result   4
  1.3   What is new?   6
  1.4   Challenges   8
      1.4.1   Varying friction: varying convexity of dissipation Ψ   9
      1.4.2   Making and losing contact: discontinuous dissipation   12
      1.4.3   Curved boundary: non-convex domains   14

2 The mechanical and mathematical model   **16**
  2.1   Mechanical model - a force balance law   16
      2.1.1   Neglecting inertial forces   17
      2.1.2   Elastic forces   19
      2.1.3   Dissipational forces   19
      2.1.4   Constraint forces   23
  2.2   Mathematical models   24
      2.2.1   Differential inclusion   24
      2.2.2   Evolutionary variational inequality   25
      2.2.3   Energetic formulation   25
  2.3   Stability of the quasi-static path   30
      2.3.1   Model and definition of stability   31
      2.3.2   Variation of the derivative of the quasi-static path   34
      2.3.3   Stability of the quasi-static path   38

3 Varying friction - basic mathematical strategy   **42**
  3.1   Overview   42
  3.2   Energetic formulation of the problem   43

3.3   Assumptions and existence result 43

3.4   Approximative solutions 45

3.5   Convergence 49

    3.5.1   Arzela-Ascoli 49

    3.5.2   Recursive estimates for solutions 50

    3.5.3   Lipschitz continuity & convergence 53

3.6   Existence 55

    3.6.1   Initial condition 56

    3.6.2   Stability 56

    3.6.3   Energy equality 57

4   Making and losing contact (one particle)   62

4.1   Overview 62

4.2   Assumptions, problem and result 63

4.3   Decomposition into two subproblems 66

    4.3.1   Friction-free subproblem 66

    4.3.2   Contact subproblem (under strong convexity assumption) . . 69

    4.3.3   Merging the two solutions 73

4.4   Contact subproblem under weak convexity assumption 77

    4.4.1   Increasing the convexity: a simplified problem 81

    4.4.2   Solving simplified problems 88

    4.4.3   Solving the contact subproblem 100

4.5   Examples 103

5   Making and losing contact (N particles)   108

5.1   Overview 108

5.2   Assumptions, problem and result 109

5.3   Decreasing the friction: a simplified problem 114

5.4   Solving simplified problems 117

5.5   Solving the problem 128

6   Curved boundary - the full problem   132

6.1   Overview 132

6.2   Modeling a curved boundary 133

6.3   Assumptions, problem and result 135

6.4   Transformation to a problem with flat boundary 139

6.5   Retranslating the convexity assumption 151

6.6   Special case $N = 1$ 155

A   Subdifferential calculus   160

Contents

Bibliography **162**

Lebenslauf **169**

Chapter 1

Introduction

In this thesis we prove local existence for a quasi-static (rate-independent) contact problem of a non-linear elastic system with finitely many degrees of freedom. The system may get into contact with a rigid obstacle which may have a curved boundary. Friction is modeled as dry friction according to the Coulomb law. The friction coefficient and the normal force vary with respect to the position of the system which leads to a non-associative flow rule. The problem is formulated as a differential inclusion but solved using an energetic formulation introduced by Mielke & Theil [MiT99].

To motivate the problem imagine an elastic system which is determined by the position of N particles, see Figure 1.1. Assume this system is placed in front of you on your desk and that you push it slowly with your finger. The first thing you will

Figure 1.1: Elastic system sliding over a desk

observe is that the system deforms a bit due to elasticity. If you keep on pushing the system it will start to slide and we face friction. The coupling between elasticity and friction makes it hard to predict the movement of the system mathematically. The goal of this work is to predict the movement of the system. We are particularly interested in the influence of the geometry of the obstacle or desk. In contrast to earlier approaches, we take into account the full geometric non-linearity which arises from a curved boundary and do not linearize it or simplify it.

To keep this task simple we push the elastic system only very slowly with our finger. Then we hope that the system reacts slowly in time and thus has a small velocity. Since the kinetic energy is proportional to the velocity in square this would lead to a very small kinetic energy. This motivates our simplification: we neglect the kinetic energy in our models. In the literature such models are often called quasi-static models. Since we do not consider any viscous effects our models will be rate-independent. This means that the principal behavior of the system is independent of the rate of loading, e.g. if the finger acts exactly half as fast then the system responds only half as fast. But even if the velocity of the system is proportional to the rate of loading and we assume this rate of loading to be small, there remains the question whether the system responds with a small or finite velocity or whether it produces jumps. It turns out that this physical reasoning is also the key for the mathematical existence proof. Thus the central question of this work is:

When does the elastic system have a small or finite velocity?

Since we allow for curved boundaries we could make this question more precise and ask:

Which geometry of the boundary supports small velocities of the system?

For example have a look at the following two pictures. Does a convex or a concave boundary reduce the velocity of the system? The answer to this question is

convex   concave

Figure 1.2: Which geometry supports small velocities?

intuitively clear on physical grounds but the challenge is to derive the answer in a precise and rigorous way from the mathematical model.

1.1 Modeling

We now present the model to be considered in a simplified way. Let $N \in \mathbb{N}$ be the number of particles. The state of the system is described by $z \in \mathbb{R}^{3N}$ and for $j = 1, \ldots, N$ the vector $z^{(j)} := \begin{pmatrix} z_{3j-2} & z_{3j-1} & z_{3j} \end{pmatrix}^{\top} \in \mathbb{R}^3$ denotes the position of a single particle. The fact that the particles cannot penetrate the rigid desk is modeled via the condition $z^{(j)} \in \mathcal{S}$ for some non-empty set $\mathcal{S} \subset \mathbb{R}^3$. For example, in the case of a flat boundary we put $\mathcal{S} := \{z \in \mathbb{R}^3 : z_3 \geq 0\}$. The admissible set for the whole system is defined via $\mathcal{A} := \{z \in \mathbb{R}^{3N} : z^{(j)} \in \mathcal{S}, j = 1, \ldots, N\}$.

We want to predict the evolution of the elastic system during some given time interval $[0, T]$ with $T > 0$ and starting in $Z_0 \in \mathcal{A}$. We are thus interested in a function $z \in W^{1,\infty}([0, T], \mathcal{A})$. The mathematical choice of the space of Lipschitz continuous functions as function space coincides with the physical idea of looking for solutions with bounded velocities. This is necessary if we want to neglect kinetic energies.

We will model the problem by its energies and denote by $\mathcal{E}(t, z)$ the elastic energy of the elastic system. Note that the energy depends on time and the position of the particles only and not on the velocity. Thus, as announced, we do not take into account kinetic energy.

Also friction is modeled on an energetic level by a dissipation functional $\Psi(t, z, v)$ which depends on time, position of the particles and a sliding vector $v \in \mathbb{R}^{3N}$. Roughly speaking the functional $\Psi(t, z, y - z)$ describes how much energy is dissipated due to friction if at a time t the system is in state z and then slides instantaneously to the state $y \in \mathcal{A}$. More precisely we model friction following the Coulomb friction law and our dissipation functional will have the structure $\Psi(t, z, v) = \sum_{j=1}^{N} \left(-D_{z^{(j)}}\mathcal{E}(t, z)\nu(z^{(j)}) \right)_+ \mu(z^{(j)}) \|v^{(j)}\|$. Here $\nu \in \mathbb{R}^3$ is the outward unit normal vector of the obstacle such that $\left(D_{z^{(j)}}\mathcal{E}(t, z)\nu(z^{(j)}) \right)_+$ describes the (positive) normal force with which the j-th particle is pressed onto the boundary while μ is the coefficient of friction.

The precise mathematical problem can be formulated in three equivalent ways.

Problem 1.1 *Find $\Delta \in (0, T]$ and $z \in W^{1,\infty}([0, \Delta], \mathcal{A})$ such that $z(0) = Z_0$ and for almost all $t \in [0, \Delta]$ we have*

$$0 \in D\mathcal{E}(t, z(t)) + \partial_v \Psi(t, z(t), \dot{z}(t)) + \mathcal{N}_{\mathcal{A}}(z(t)) \quad \subset \mathbb{R}^{3N^*}. \tag{DI}$$

From a physical point of view this differential inclusion is a force balance law and thus formulated in the dual space $\mathbb{R}^{3N^*} = \mathbb{R}^{1 \times 3N}$ of the state space $\mathbb{R}^{3N}$. Here $D\mathcal{E} = \frac{\partial}{\partial z}\mathcal{E}$ are the elastic forces. The possible frictional forces are expressed by the subdifferential $\partial_v \Psi$ which is a set-valued functional. The normal cone $\mathcal{N}_{\mathcal{A}}$ of the admissible set $\mathcal{A}$ in $z(t)$ represents the constraint forces that prevent the particles from leaving the set $\mathcal{A}$.

An equivalent reformulation of the problem is the following quasi-variational inequality.

Problem 1.2 *Find $\Delta \in (0, T]$ and $z \in W^{1,\infty}([0, \Delta], \mathcal{A})$ such that $z(0) = Z_0$ and for almost all $t \in [0, \Delta]$ we have*

$$0 \leq D\mathcal{E}(t, z(t))(v - \dot{z}(t)) + \Psi(t, z(t), v) - \Psi(t, z(t), \dot{z}(t)) \quad \text{for all } v \in \mathcal{T}_{\mathcal{A}}(z(t)). \tag{VI}$$

Here $\mathcal{T}_{\mathcal{A}}(z)$ denotes the tangential cone of $\mathcal{A}$ in z.

The following energetic problem formulation will be the most convenient formulation to prove existence results.

Problem 1.3 *Find* $\Delta \in (0, T]$ *and* $z \in \mathrm{W}^{1,\infty}([0,\Delta], \mathcal{A})$ *such that* $z(0) = Z_0$ *and for all* $t \in [0, \Delta]$ *the following two conditions hold:*

$$\mathcal{E}(t, z(t)) \leq \mathcal{E}(t, y) + \Psi(t, z(t), y - z(t)) \quad \text{for all } y \in \mathcal{A} \tag{S}$$

and $$\mathcal{E}(t, z(t)) + \int_0^t \Psi(\tau, z(\tau), \dot{z}(\tau)) \mathrm{d}\tau = \mathcal{E}(0, z(0)) + \int_0^t \partial_t \mathcal{E}(\tau, z(\tau)) \mathrm{d}\tau. \tag{E}$$

Condition (S) ensures that the system is energetically stable at each time $t \in [0, T]$ in the following sense. At time t our system has the energy $\mathcal{E}(t, z(t))$. Switching to another admissible state $y \in \mathcal{A}$ might lead to a release of energy because we might have $\mathcal{E}(t, y) < \mathcal{E}(t, z(t))$. But (S) guarantees that the loss of energy due to friction is higher and in such a sense our solution $z(t)$ has to be energetically stable.

The energy equality (E) represents energy conservation. The right-hand side represents the energy we have at time $t = 0$ plus the energy that is put into the system up to time t by the work of external forces, here the finger. These energies equal the energies on the left hand side which describe the energy that we have at time t plus the energy the system lost due to friction until time t.

The equivalence of these three formulations is discussed in Chapter 2.

1.2   Assumptions and main result

We now present the most important assumptions for existence of a solution and, as we will see, equivalently for smallness of the velocity. For the energy functional $\mathcal{E}$ this is uniform convexity in some neighborhood of the initial time $t = 0$ and state $Z_0 \in \mathcal{A}$. For this we denote by $\mathrm{H}(t, z) = \frac{\partial^2}{\partial z^2}\mathcal{E}(t, z) \in \mathbb{R}^{3N \times 3N}$ the Hessian matrix of $\mathcal{E}$ with respect to z. We assume that there exists a time span $\Delta \in (0, T]$, radius $r > 0$ and $\alpha > 0$ such that

$$v^\top \mathrm{H}(t, z)v \geq \alpha \|v\|^2 \quad \text{for all } v \in \mathbb{R}^{3N}, t \in [0, \Delta] \text{ and } z \in \mathcal{A} \cap \mathcal{B}_r(Z_0).$$

In the case of a purely linear model of elasticity the energy functional would be quadratic and we would have $v^\top \mathrm{H}(t, z)v = v^\top \mathrm{H}_0 v$. Thus, we are far more general and allow for non-linear elasticity but we keep convexity.

We will see that also the dissipation functional is convex with respect to its third variable. Lost energy is proportional to the product of frictional force and sliding distance. According to the Coulomb friction law we model the frictional force as the product of roughness μ and normal force $\sigma^{(j)}(t,z)$. Here $\mu : \partial\mathcal{S} \to [0,\infty)$ is the coefficient of friction which models the roughness of the boundary. For $z^{(j)} \in \partial\mathcal{S}$ we denote by $\nu(z^{(j)}) \in \mathbb{R}^3$ the outer normal vector of $\mathcal{S}$. Hence we define the normal force $\sigma^{(j)}(t,z)$ with which the j-th particle $z^{(j)}$ is pressed onto the boundary as the pairing $\sigma(t,z) := -\partial_{z^{(j)}}\mathcal{E}(t,z)\nu(z^{(j)}) \in \mathbb{R}$.

Thus, the energy which is dissipated by a single particle $z^{(j)}$ sliding along the vector $v^{(j)} \in \mathbb{R}^3$ reads

$$\Psi^{(j)}(t,z,v^{(j)}) := \begin{cases} \mu(z^{(j)})\sigma^{(j)}(t,z)_+ \|v^{(j)}\| & \text{for} \quad z^{(j)} \in \partial\mathcal{S}, \\ 0 & \text{for} \quad z^{(j)} \in \mathrm{int}\mathcal{S}. \end{cases}$$

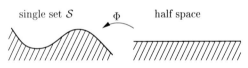

The full dissipation functional is then described by $\Psi(t,z,v) = \sum_{j=1}^{N} \Psi^{(j)}(t,z,v^{(j)})$. The index '+' in the definition of $\Psi^{(j)}$ indicates that we only take positive normal forces into account, i.e. $\sigma_+ = \max\{\sigma, 0\}$. The major assumption on the dissipation is the existence of a local Lipschitz constant $q > 0$ for the frictional forces, i.e. there exist $\Delta \in (0,T]$ and $r > 0$ such that $\forall t \in [0,\Delta], \forall z_1, z_2 \in \mathcal{A} \cap \mathcal{B}_r(Z_0)$ we have

$$\left| \mu(z_2^{(j)})\sigma^{(j)}(t,z_2)_+ - \mu(z_1^{(j)})\sigma^{(j)}(t,z_1)_+ \right| \leq q\|z_2 - z_1\|.$$

This important assumption is the reason why we can construct local existence results only. In fact uniform convexity of $\mathcal{E}$ and global Lipschitz continuity of $\mu(z^{(j)})\sigma^{(j)}(t,z)$ exclude each other on unbounded domains and arbitrary time intervals.

To model the geometry of the boundary we assume for a moment that there exists a transformation $\Phi \in \mathrm{C}^2(\mathbb{R}^3, \mathbb{R}^3)$ which locally maps the half space $\{z \in \mathbb{R}^3 : z_3 \geq 0\}$ on our single admissible set $\mathcal{S} \subset \mathbb{R}^3$ with curved boundary. The curvature or convexity of the boundary can be calculated via the second derivative of Φ, i.e. the matrix $\mathrm{D}^2\Phi(z) \in \mathbb{R}^{3\times3}$. After these assumptions on the functionals $\mathcal{E}$ and Ψ and on the transformation Φ we introduce the most crucial assumption which establishes the link between elasticity, friction and geometry. Using a rather symbolic notation our existence result will be of the following type.

Theorem 1.4 (Existence) *Let the above assumptions hold and assume*

$$\alpha + \sigma(0, Z_0)_+ \mathrm{D}^2\Phi(0) > q \tag{1.1}$$

then there exists $\Delta \in (0,T]$ and a solution $z \in \mathrm{W}^{1,\infty}([0,T], \mathcal{A})$ of Problem 1.1.

The detailed and main existence theorems of this thesis will be given in Chapter 6, namely Theorem 6.5 and Theorem 6.19. We briefly discuss the crucial Assumption

(1.1) for the case of a single particle, i.e. $N = 1$. For a flat boundary we have $D^2\Phi \equiv 0$ and thus assume $\alpha > q$. Hence the convexity of the elastic energy has to dominate the Lipschitz constant of the dissipational forces. The geometry of a curved boundary, i.e. $D^2\Phi \not\equiv 0$, only influences the convexity of the energy and matters only if the particle is pressed onto the boundary, i.e. $\sigma_+ > 0$. The more the particle is pressed onto the boundary the more the geometry matters. A convex geometry, i.e. $D^2\Phi > 0$, 'increases' the convexity while a concave boundary, i.e. $D^2\Phi < 0$, 'decreases' the convexity. See, also Figure 1.3. It is intuitively clear that

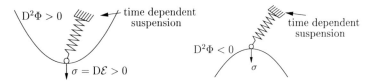

Figure 1.3: Influence of the geometry of the boundary

a convex geometry stabilizes or slows down a particle while a concave geometry may produce jumps, especially if the particle is pressed onto the obstacle by a strong normal force.

Our existence results are local and thus restricted to small time spans $\Delta > 0$. As in the theory of ordinary differential equations we can extend the results to larger time spans until either we reach the final time T or our major assumption $(\alpha + \sigma_+ D^2\Phi) >$ q does not hold any longer, i.e. we have $\alpha(t, z(t)) + \sigma(t, z(t))_+ D^2\Phi(z(t)) - q(t, z(t)) \to$ 0 for $t \to t_*$. In Section 4.5, i.e. for $D^2\Phi = 0$, we present examples which show that for $\alpha = q$ jumps may occur and no solution exists. Hence, our crucial assumption is sharp.

The question of uniqueness is not addressed in this work. Ballard presents in [Bal99] a counter-example to uniqueness for a one particle problem which is included in our problem formulations even if the friction coefficient is arbitrary small. This indicates that uniqueness should not be expected for our models under consideration.

1.3 What is new?

Interest in the question of existence for quasi-static contact problems in linear elasticity with dry friction reached a peak in the literature in the nineties of the last century. The discussion started with the first existence results for the continuous problem of Klarbring, Mikelić and Shillor [KMS88, KMS89, KMS91]. They still considered regularized problems. Interest then waned towards the end of the nineties. It was Andersson [And00] and Rocca [Roc99, Roc01] who solved the com-

plete model on the continuous level. Andersson [And99] also solved the problem on
the discrete level for N particles. In the last years this problem found clearly less
attention. Touzaline at al. still work on the continuous model and have extended
it to non-linear elastic bodies in a series of articles [Tou06b, Tou06a, ToT07]. But
most authors have started to expand their models principally including thermal
effects, see [EcJ01, AK*02, ChA04, AK*05]. Also the energetic approach which we
use as a method has been expanded by our colleagues in [MiP07, AMS08] such that
thermal effects can be included. Other expansions include non-coercivity of the
elastic energy [AnR06], wear [SST04] and damage [HSS01].

So why reconsider a problem which has been considered as solved for several years
- and this even on the continuous level!? It is the geometry of the boundary which
attracted our attention. All analytical results up to now assume either a simple
planar geometry of the obstacle or take it into account only in a very simplified
way. All this is justifiable for small deformations or displacements only. But what
happens if the obstacle boundary is curved and for finite displacements, thus in non-
linear theory? This question was already addressed by Martins et al. in [PiM03] but
not completely resolved. The principal achievement of this work is that we are able
to determine the precise influence of the geometry of the obstacle boundary $\partial \mathcal{A} \in$
C^3 on the question of existence for quasi-static and finite dimensional problems.
The most important results are thus Theorem 6.5 and Theorem 6.19. Even if our
existence results are limited to the finite-dimensional setting we hope to give a
hint and some insight what should be expected on the continuous level, e.g. in
finite-strain elasticity. In Figure 1.4 we illustrate the classical simplified unilateral
constraint on the continuous level and why it should be replaced by a geometrically
exact constraint in case of non-linear elasticity. The boundary of the obstacle is

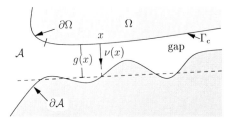

Figure 1.4: Simplified unilateral constraint indicated by dotted line

denoted by $\partial \mathcal{A}$. Let $\Omega \subset \mathbb{R}^3$ be the elastic body and let $u(t, \cdot) : \Omega \to \mathbb{R}^3$ denote
the displacement. The geometrically exact constraint then reads $x + u(t, x) \in \mathcal{A}$
for all $t \in [0, T]$ and $x \in \partial \Omega$. Obviously this constraint is in general non-linear
and non-convex. In contrast the simplified constraint which is used in the literature

is illustrated in Figure 1.4. For $x \in \Gamma_c$ we define the gap $g(x) \geq 0$ such that $x + g(x)\nu(x) \in \partial\mathcal{A}$ holds with the outward normal vector $\nu(x) \in \mathbb{S}^2$. The simplified constraint reads $\langle u(x), \nu(x) \rangle \leq g(x)$ for all $x \in \Gamma_c$. Thus the constraint does not take tangential displacement into account and is equivalent in assuming $x + u(t, x)$ to remain above the dotted line for all $t \in [0, T]$ and our specific $x \in \Gamma_c$. Hence this constraint is only a good approximation if the boundaries $\partial\Omega$ and $\partial\mathcal{A}$ are locally planar and parallel and displacements are small.

1.4   Challenges

In this work we face three major mathematical difficulties. These difficulties and the ideas how to solve them determine the structure of the thesis, see Chapter 3–6. Each mathematical difficulty is due to some physical phenomenom which we illustrate in the following. First the frictional force $\mu(z^{(j)})\sigma^{(j)}(t, z)$ changes

Figure 1.5: Varying frictional force: varying convexity of dissipation Ψ

with respect to z which may lead to instabilities. In Figure 1.5 the particle jumps because the roughness μ changes from stone to ice. We want to emphasize that this jump has nothing to do with an abrupt change of roughness. We even get jumps for $\mu \in C^\infty(\partial\mathcal{A}, [0, \infty))$, see the examples in Section 4.5. We present sufficient conditions preventing jumps in Chapter 3.

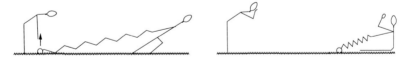

Figure 1.6: Contact vs. non-contact: discontinuous dissipation Ψ

In Chapter 4 and 5 we additionally take the second challenge into account. This challenge is due to making and losing contact, see Figure 1.6. Mathematically this leads us to a discontinuous dissipation functional even if we assume smooth frictional forces. In fact this will be the most severe difficulty of this work.
In Chapter 6 the third and last difficulty is taken into account. The geometric non-linearity, see Figure 1.7, forces us to deal with non-convex admissible sets $\mathcal{A}$.

Figure 1.7: Non-linear geometry: non-convex set $\mathcal{A}$

1.4.1 Varying friction: varying convexity of dissipation Ψ

In Chapter 3 we deal with the difficulty of varying frictional forces. Here we sketch the principal ideas, by keeping the situation simple, avoiding the other two difficulties. For this we assume a convex boundary and that no switching between contact and non-contact occurs. This means that during the whole process each single particle either is in permanent contact, i.e. $z^{(j)} \in \partial \mathcal{S}$, or is never in contact, i.e. $z^{(j)} \in \mathcal{S}/\partial \mathcal{S}$. This implies that either $\Psi^{(j)}(t, z, v^{(j)}) = \mu(z^{(j)})\sigma^{(j)}(t, z))_+ \|v^{(j)}\|$ or $\Psi^{(j)} \equiv 0$. Thus the dissipation functional $\Psi = \sum_{j=1}^{N} \Psi^{(j)}$ is continuous with respect to z and there at least locally exists some constant $q > 0$ such that

$$|\Psi(t, z, v) - \Psi(t, \tilde{z}, v)| \leq q\|z - \tilde{z}\|\|v\|. \tag{1.2}$$

Further we assume the set $\mathcal{A} \subset \mathbb{R}^{3N}$ to be convex. The existence proof consists of three steps.

1. **Approximation**

   The advantage of the (S)&(E) formulation is that there exists a rather simple discrete approximating scheme, see [MiT99, Mie05]. We split the time interval $[0, T]$ uniformly into discrete time points:

   $$0 = t_0 < t_1 < \cdots < t_N = T$$

   with fineness $f_{\Pi} := t_k - t_{k-1}$. We define the corresponding values incrementally via

   $$z_k \in \operatorname{argmin}\left\{\mathcal{E}(t_k, y) + \Psi(t_{k-1}, z_{k-1}, y - z_{k-1}) : y \in \mathcal{A}\right\}. \tag{IP}$$

   Hence, the choice of z_k at time t_k is such that not only the elastic energy is kept small but also the energy we need due to friction in order to attain this new position z_k is little. Starting from the previous position z_{k-1} at time t_{k-1} this second energy is roughly expressed by the dissipation functional in (IP). Since both functionals were assumed to be convex and $\mathcal{E}$ is uniformly convex with respect to z and since the set $\mathcal{A}$ is convex there exists a unique solution z_k of (IP). Hence convexity plays an important role in all our models.

2. **Convergence**

   The difficult part will always be the convergence part. Here most of the new ideas can be found. Let us assume the fineness to satisfy $f_\Pi^{(n)} \to 0$ for $n \to \infty$. Then we have to show that

$$z^{(n)} \overset{*}{\rightharpoonup} z \in W^{1,\infty}([0,T], \mathcal{A})$$

for piecewise linear approximations $z^{(n)} \in W^{1,\infty}([0,T], \mathcal{A})$ constructed from the discrete solutions of (IP).

3. **Existence**

   In this step we have to show that the limit function of the previous step supplies us with a solution of the (S)&(E)-Problem. This step is nowadays well understood and was carried out in many articles dealing with the energetic problem formulation, see for example [Mie05] or [MiR07]. In our problem there was nothing really new to add and we followed the ideas of the above articles.

Hence, the main task is Step 2 where we have to establish convergence. This is always the most difficult part in the existence proofs for energetic problems. Since in this thesis we deal with a finite dimensional problem we use the Arzela-Ascoli convergence theorem. Thus it is sufficient to establish the existence of a Lipschitz constant $c_{\mathrm{lip}} > 0$ independent of the fineness of the partition $f_\Pi^{(n)}$ (and thus of n) such that

$$\|z_{k+1}^{(n)} - z_k^{(n)}\| \le c_{\mathrm{lip}} |t_{k+1}^{(n)} - t_k^{(n)}| \tag{1.3}$$

holds for the discrete solutions of the incremental problem (IP). In other words, we obtain existence of the mathematical problem if the approximative solutions have a uniformly bounded velocity. We want to recall that bounded velocities are necessary for our model to be physically reasonable since we neglect kinetic energies. Thus the mathematically sufficient criteria is physically necessary and they match nicely. In the whole work the construction of the above Lipschitz estimate is the central step. We will find that convexity is not only good to define the single values z_k but also to show that with respect to time they do not shift too much. For simplicity, we do not take into account dissipation for a moment and define

$$z_k \in \operatorname{argmin} \{\mathcal{E}(t_k, y) \, : \, y \in \mathcal{A}\} . \qquad \text{(no dissipation)}$$

The corresponding picture looks as follows:

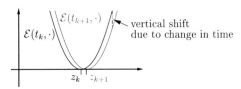

Due to the change in time from t_k to t_{k+1} the values change in vertical direction. Due to the high convexity the graph consequently shifts only little in horizontal direction. In the next picture we have the same vertical shift (dotted line) but due to less convexity this induces a bigger horizontal shift. Thus a high convexity of $\mathcal{E}$ generates small differences of minimizers.

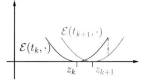

For the corresponding mathematical calculus we use that z_k is the minimizer at time t_k and $\mathcal{E}$ to be quadratically bounded from below, i.e. $v^\top H(t,z)v \geq \alpha \|v\|^2$, to find
$$\frac{1}{2}\alpha\|z_{k+1} - z_k\|^2 \leq \mathcal{E}(t_k, z_{k+1}) - \mathcal{E}(t_k, z_k).$$
Deriving the analogue estimate for time t_{k+1} and minimizer z_{k+1} and adding up both estimates we deduce
$$\alpha\|z_{k+1} - z_k\|^2 \leq \mathcal{E}(t_k, z_{k+1}) - \mathcal{E}(t_k, z_k) + \mathcal{E}(t_{k+1}, z_k) - \mathcal{E}(t_{k+1}, z_{k+1})$$
$$\leq c\|z_{k+1} - z_k\|\,\|t_{k+1} - t_k\|.$$

After division with $\alpha\|z_{k+1} - z_k\|$ we exactly find our desired estimate (1.3) with the Lipschitz constant $c_{\text{lip}} = c/\alpha$, i.e.
$$\|z_{k+1} - z_k\| \leq \frac{c}{\alpha}|t_{k+1} - t_k|.$$

If we now recall our full incremental problem (IP) with dissipation functional $\Psi(t_{k-1}, z_{k-1}, y - z_{k-1})$ then the same procedure holds apart from the fact that the convex shape of $\Psi(t_{k-1}, z_{k-1}, \cdot)$ depends not only on time but also on z. Thus

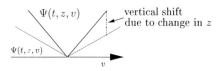

Figure 1.8: Vertical shift of Ψ to change in z

there is another vertical shift induced by a difference in z which we control using the estimate (1.2). Then we find for some $c_* \geq c$
$$\|z_{k+1} - z_k\| \leq \frac{c_*}{\alpha}|t_{k+1} - t_k| + \frac{q}{\alpha}\|z_k - z_{k-1}\|.$$

The additional term $\|z_k - z_{k-1}\|$ on the right hand side we do not want. But lowering the index by one in the above estimate and applying it recursively we find a geometric sum with increasing powers of the factor $\frac{q}{\alpha}$. Now the crucial assumption in our existence theorem reads $\alpha > q$ or equivalently $\frac{q}{\alpha} < 1$ and we will find $c_{\text{lip}} = \frac{c}{\alpha - q}$.

1.4.2   Making and losing contact: discontinuous dissipation

In Chapter 4 and 5 we still assume $\mathcal{A}$ to be convex but allow the particles $z^{(j)}$ to switch between contact and non-contact. Thus the dissipation functional is discontinuous and defined by $\Psi^{(j)}(t, z, v^{(j)}) = \mu(z^{(j)})\sigma^{(j)}(t, z)_+ \|v\|$ for $z^{(j)} \in \partial \mathcal{S}$ and $\Psi^{(j)}(t, z, v) = 0$ for $z^{(j)} \in \text{int}\mathcal{S}$. We start with the question:

What is the difficulty with the discontinuity of the single dissipations?

The difficulty is that we want to follow the incremental scheme proposed in the previous subsection. There we controlled the vertical shift induced by changes in z using an estimate of the type

$$|\Psi(t, z, v) - \Psi(t, \tilde{z}, v)| \leq q\|z - \tilde{z}\|\|v\|,$$

see also Figure 1.8. Without continuity of $\Psi^{(j)}$ with respect to z such an estimate will not hold.

Which is the basic observation?

If a particle $z^{(j)}$ is making or losing contact at time t then our physical intuition tells us that $\sigma^{(j)}(t, z(t)) = 0$ holds. We have no normal force. As a consequence we have no friction and though having contact $z^{(j)}(t) \in \partial \mathcal{S}$ we have $\Psi^{(j)}(t, z(t), \cdot) = 0$. Thus at a physical reasonable switch between contact and non-contact the dissipation behaves continuously.

Which trick do we deduce from this observation?

There are two different tricks possible. The more elegant one leads us to a quite weak assumption on the correlation between elasticity and friction. Unfortunately this trick works only for a single particle, i.e. the case $N = 1$, see Chapter 4.

The principal idea is that we avoid the discontinuity by considering two auxiliary problems. In the first problem we put $\Psi \equiv 0$ and call the solution z_f friction-free. Note that due to the stability condition (S) this solution is uniquely defined by $\mathcal{E}(t, z_f(t)) \leq \mathcal{E}(t, y)$ for all $y \in \mathcal{A}$. In the second auxiliary problem we constrain the particle to the boundary. Thus the dissipation is continuously defined by $\Psi(t, z, v) =$

$\mu(z)\sigma(t,z)_+ \|v\|$. We call the solution $z_c : [0,T] \to \partial\mathcal{A}$ the constrained solution. Based on our above observation we then make the ansatz

$$z(t) := \begin{cases} z_c(t) & \text{if } \sigma(t, z_c(t)) > 0, \\ z_f(t) & \text{else.} \end{cases}$$

Why does this trick work? Whenever the constrained solution z_c is pressed onto the boundary it is a physically reasonable solution. When z_c would like to leave the boundary we have $\sigma(t, z_c(t)) = 0 = \Psi(t, z_c(t), \cdot)$. Being free of friction it coincides with the unique friction-free solution z_f which is the right solution for being out of contact. Unfortunately this trick does not work for more than one particle due to the mutual interaction between particles, see Example 4.14

In the N particle case we need stronger assumptions. The reasons are quite technical but one of the most important reasons is that we are not able to control changes of directions in the N particle case.

Thus, to be able to treat the general case $N \in \mathbb{N}$ we proceed in a different way and use stronger assumptions and a more powerful trick, see Chapter 5. This time we do not solve auxiliary problems but introduce a parameter $\gamma > 0$ in the approximating scheme. We replace the single dissipation functionals $\Psi^{(j)}$ by new dissipation functionals

$$\Psi_\gamma^{(j)}(t, z, v) := \begin{cases} \mu(z^{(j)})\left(\sigma^{(j)}(t, z) - \gamma\right)_+ \|v^{(j)}\| & \text{if } z_{3j} = 0, \\ 0 & \text{if } z_{3j} > 0. \end{cases}$$

Hence we replace in the dissipation functional the normal force σ_+ by the shifted normal force $(\sigma - \gamma)_+$, see Figure 1.9. The consequence is that these new dissipation

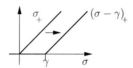

Figure 1.9: The γ-trick

functionals are continuous (equal zero) for small normal forces $\sigma(t, z) < \gamma$. What is the idea behind this trick? Physical solutions have zero normal force when they make and lose contact. Approximative solutions unfortunately do not match this behavior exactly. But good approximations (i.e. for small fineness f_Π) have small normal forces (i.e. $\sigma^{(j)}(t_k, z_k) < \gamma$) while making and losing contact. Thus the above new dissipations $\Psi_\gamma^{(j)}$ are again continuous with respect to good approximations. We will pass to the limit as follows. First we let the step size or fineness of the partitions converge to zero, i.e. $f_\Pi \to 0$ while $\gamma > 0$ is fixed. Afterwards we consider $\gamma \to 0$.

1.4.3   Curved boundary: non-convex domains

After Chapter 5 we are able to handle varying frictional forces and of the switching between contact and non-contact. In Chapter 6 we then additionally take curved boundaries into account and consider the full problem. What is the difficulty with curved boundaries? Curved boundaries lead to non-convex admissible sets $\mathcal{A}$, but we still want to use the incremental approach with its incremental problem

$$z_k \in \operatorname{argmin} \left\{ \mathcal{E}(t_k, y) + \Psi(t_{k-1}, z_{k-1}, y - z_{k-1}) \ : \ y \in \mathcal{A} \right\}. \tag{IP}$$

For non-convex sets $\mathcal{A}$ we are not able to construct (unique) solutions and to apply our existence results of Chapter 4 and 5. The idea is to equivalently reformulate the curved problem as a flat problem. For this we introduce a transformation $\Phi \in C^2(\bar{\mathcal{A}}, \mathcal{A})$, see Figure 1.10. We replace the energy functional $\mathcal{E}$ and the dissipation

Curved boundary Φ Flat boundary

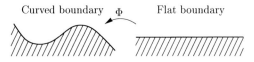

Figure 1.10: The transformation

functional Ψ by $\bar{\mathcal{E}}(t, \bar{z}) := \mathcal{E}(t, \Phi(\bar{z}))$ and $\bar{\Psi}(t, \bar{z}, v) := \Psi(t, \Phi(\bar{z}), \mathrm{D}\Phi(\bar{z})v)$. These functionals are defined with respect to the flat boundary. The major task will be to reformulate the existence assumptions on $\bar{\mathcal{E}}$ and $\bar{\Psi}$ for the flat problem as assumptions on $\mathcal{E}, \Psi$ and $\mathcal{A}$.

Chapter 2

The mechanical and mathematical model

In Section 2.1 we present the mechanical model and discuss the physical background of our problem. We introduce the precise mathematical formulation in Section 2.2. Indeed, we present several equivalent mathematical formulations hoping that each reader finds a formulation he is familiar with. Since the problem under consideration is a simplification we give in Section 2.3 a mathematical proof that under certain circumstances the solutions of the simplified and full problem remain close to each other.

2.1 Mechanical model - a force balance law

To introduce the mechanical model we consider an elastic system of N particles which might make contact with a rigid obstacle. See also Figure 2.1 below. By

$$z \in \mathbb{R}^{3N}$$

we describe the state or position of the system with N particles. For $j = 1, \ldots, N$ we denote the position of a single particle by

$$z^{(j)} := \begin{pmatrix} z_{3j-2} & z_{3j-1} & z_{3j} \end{pmatrix}^{\top}.$$

The fact that each single particle cannot penetrate the rigid obstacle is modeled by the restriction $z^{(j)} \in \mathcal{S}$ with $\mathcal{S} \subset \mathbb{R}^3$ denoting the admissible set. For the time being we assume for simplicity

$$\mathcal{S} = \left\{ y \in \mathbb{R}^3 \ : \ y_3 \geq 0 \right\}.$$

Thus our obstacle has a simple and flat boundary $\partial \mathcal{S} := \{ y \in \mathbb{R}^3 \ : \ y_3 = 0 \}$ and for $z \in \partial \mathcal{S}$ the outward normal vector is described via

$$\nu(z) := \begin{pmatrix} 0 & 0 & -1 \end{pmatrix}^{\top}.$$

Later on we study more general geometries. The modeling and description of a curved boundary is presented in Chapter 6. In the present case the admissible set of the whole system has the form $\mathcal{A} = \left\{ z \in \mathbb{R}^{3N} \,:\, z^{(j)} \in \mathcal{S} \right\}$. See, also Figure 2.1 where we present an elastic system with five particles. For simplicity the picture is two dimensional only. Since the system is time-dependent we fix a time $T > 0$ and

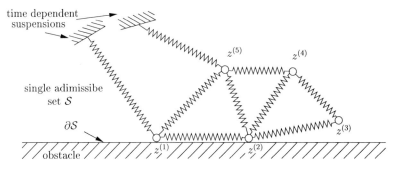

Figure 2.1: The N particle system

seek for a solution $z : [0, T] \to \mathcal{A}$ starting at some prescribed initial state $Z_0 \in \mathcal{A}$. The evolution of the solution is modeled by assuming a force balance to hold for each time $t \in [0, T]$. Using a symbolical notation we express this force balance law via

$$0 \in \left(\underbrace{m\ddot{z}(t)}_{-\text{inertial}} \right) + \underbrace{e(t, z(t))}_{-\text{elastic}} + \underbrace{\mathcal{D}(t, z(t), \dot{z}(t))}_{-\text{frictional}} + \underbrace{\mathcal{C}(z(t))}_{-\text{constraint}} \quad \subset \mathbb{R}^{3N^*}.$$

(force balance law)

Thus, we take into account four different forces: inertial forces which we will neglect and for this put above only in brackets, elastic forces which correspond to the elasticity of the system, frictional forces in case of particles being in contact and constraint forces which prevent the particles from penetrating the obstacle. All forces belong to the dual space $\mathbb{R}^{3N^*} \cong \mathbb{R}^{1 \times 3N}$ of our state space $\mathbb{R}^{3N} \cong \mathbb{R}^{3N \times 1}$. Each of the four different kinds of forces is discussed in detail in the following Subsections 2.1.1 – 2.1.4. In our convention the above forces are always the opposite of the forces acting on the particles. This explains the minus in front of the denominations.

2.1.1 Neglecting inertial forces

Let $m > 0$ be the mass of a single particle and $\ddot{z}$ denote the second derivative $\frac{\mathrm{d}^2}{\mathrm{d}t^2} z$ then the inertial force acting on the particle is opposite to the acceleration and thus described via $-m\ddot{z}(t)$. We mentioned above that there are two ways to model the

N-particle system. On the one hand there is the physically correct modeling taking
into account the inertial force:

$$0 \in m\ddot{z}(t) + \mathrm{e}(t, z(t)) + \mathcal{D}(t, z(t), \dot{z}(t)) + \mathcal{C}(z(t)) \quad \subset \mathbb{R}^{3N^*}.$$

(dynamic force balance law)

On the other hand we find a simplified model by neglecting the inertial force:

$$0 \in \mathrm{e}(t, z(t)) + \mathcal{D}(t, z(t), \dot{z}(t)) + \mathcal{C}(z(t)) \quad \subset \mathbb{R}^{3N^*}. \text{ (quasi-static force balance law)}$$

The main focus of this work is on the quasi-static model. We now give a physical
motivation for neglecting the inertial term. Our elastic system typically faces two
time-scales. An internal time scale which is usually determined by the time scale
of the eigen frequencies of the system. Our first important assumption is that this
internal time scale is small compared with the time-scale of the external loading.
The relation between the fast physical time t and the slow loading time $\tilde{t}$ is expressed
by the loading rate $1 \gg \varepsilon > 0$ via $\tilde{t} = \varepsilon t$. For example if t is expressed in seconds
and $\tilde{t}$ in years then we find $\varepsilon \approx 10^{-8}$. Thus the dynamic force balance law expressed
with respect to the new time $\tilde{t}$ reads

$$0 \in \varepsilon^2 m\ddot{z}(\tilde{t}) + \mathrm{e}(\tilde{t}, z(\tilde{t})) + \mathcal{D}(\tilde{t}, z(\tilde{t}), \varepsilon\dot{z}(\tilde{t})) + \mathcal{C}(z(\tilde{t})) \quad \subset \mathbb{R}^{3N^*}.$$

For small $\varepsilon > 0$ or equivalently for slowly evolving external forces the inertial forces
are relatively small and this motivates us to neglect them in the modeling. The
second important assumption is that the dissipational term $\mathcal{D}$ is homogenous of
degree zero with respect to the third argument: For all $t \in [0, T], z \in \mathcal{A}, v \in \mathbb{R}^{3N}$
and $\lambda > 0$ we have

$$\mathcal{D}(t, z, \lambda v) = \mathcal{D}(t, z, v).$$

Physically this means that the frictional forces principally depend on the direction
of sliding and not on the amount of the velocity, see the Coulomb friction law below
in Section 2.1.3. This important assumption does not only show that we can leave
the frictional term unchanged in the above dynamical equation but also that the
quasi-static model is rate-independent, i.e. rescaling of time in the problem leads to
the same rescaling of time for the solution. For example let z be the mathematical
solution of the quasi-static force balance with respect to time t, then $\tilde{z}$ defined via
$\tilde{z}(\tilde{t}) = z(t)$ with $\tilde{t} = \varepsilon t$ is the solution of the rescaled problem. This important
property shows that the dependence of the quasi-static solution on the time scale
is relatively simple and thus solving the quasi-static problem we can study easily
the qualitative behavior of the solution.

We present in Section 2.3 a mathematical rigorous result which states that the
solutions of dynamical problems (with inertia) remain close to solutions of simplified
quasi-static problems (without inertia) if the excitation by external forces is slow.
This result is established only for a simple one particle situation.

2.1.2   Elastic forces

The elastic behavior of the system will be modeled in a rather general way via an Energy functional $\mathcal{E} : [0, T] \times \mathcal{A} \to \mathbb{R}$. We assume that this energy functional is given as data and it describes the energy of the system depending on time t and state z of the system. The elastic force that acts on the particles is then described via

$$\mathrm{e}(t, z) = -\mathrm{D}\mathcal{E}(t, z) \in \mathbb{R}^{3N^*},$$

see also the above inclusion (DI). As for the single particles we denote for $j = 1, \ldots, N$ by $(\mathrm{D}\mathcal{E})^{(j)}(t, z) = \left(\partial_{z_{3j-2}} \mathcal{E} \quad \partial_{z_{3j-1}} \mathcal{E} \quad \partial_{z_{3j}} \mathcal{E} \right)(t, z) \in \mathbb{R}^{3^*}$ the force that corresponds to a single particle.

2.1.3   Dissipational forces

Let $z \in \mathcal{A}$ be the state of the system and fix $j \in \{1, \ldots, N\}$. We do the modeling of the single frictional forces $\mathcal{D}^{(j)} \subset \mathbb{R}^{3^*}$ for each particle $z^{(j)}$ separately and we assume $j \in \{1, \ldots, N\}$ to be fixed in the following.

Even if we are principally interested in modeling the frictional forces, we discuss in this paragraph first the energy that is lost due to friction, see (2.1). The definition of the frictional forces is then found in formula (2.5). For this we assume a particle to slide from state $z^{(j)} \in \partial \mathcal{S}$ instantaneously to the state $y \in \partial \mathcal{S}$. The dissipated energy is calculated as the sliding distance times the force, which we need to move the particle. This moving force is assumed to be proportional to the force with which the particle is pressed onto the boundary (normal force) and the roughness of the surface of the boundary. For $z \in \mathcal{A}$ with $z^{(j)} \in \partial \mathcal{S}$ we denote this normal force via

$$\sigma^{(j)}(t, z) := - \left(\mathrm{D}\mathcal{E} \right)^{(j)}(t, z) \nu(z^{(j)}).$$

The roughness is in the literature normally modeled via a scalar coefficient of friction $\mu^{(j)} : \partial \mathcal{S} \to [0, \infty)$. This leads to an isotropic model of friction. In our modeling we allow that the roughness of the surface behaves different in different sliding directions. This anisotropic behavior is modeled via matrices of friction

$$\mathrm{M}^{(j)} : \partial \mathcal{S} \to \mathbb{R}^{3 \times 3} \quad \text{with} \quad \mathrm{M}^{(j)}(z)\nu = 0 \in \mathbb{R}^3 \text{ for all } z \in \partial \mathcal{S}.$$

The dependence of $\mathrm{M}^{(j)}$ on the index $j \in \{1, \ldots, N\}$ expresses that single particles can consist of different materials and thus generate different frictional forces. The additional property $\mathrm{M}^{(j)}(z)\nu = 0 \in \mathbb{R}^3$ assures that only the tangential sliding of the particle is taken into account. We distinguish with regard to the index j between different matrices since different particles might consist of different materials and thus generate different roughness. We summarize this modeling and present the dissipation functional $\Psi^{(j)} : [0, T] \times \mathcal{A} \times \mathbb{R}^3 \to [0, \infty)$ which is defined via

$$\Psi^{(j)}(t, z, v) := \begin{cases} \sigma^{(j)}(t, z)_+ \, \|\mathrm{M}^{(j)}(z^{(j)})v\| & \text{for} \quad z^{(j)} \in \partial \mathcal{S}, \\ 0 & \text{for} \quad z^{(j)} \in \mathrm{int} \mathcal{S} \end{cases} \tag{2.1}$$

with $\beta_+ := \max\{0, \beta\}$ and the usual Euclidian norm $\|\cdot\|$. In fact in most of the work more general norms can be used except for Section 4.4. Writing σ_+ means that we take friction only into account when the particle is pressed onto the obstacle and not pulled away. Let us now assume that at time t the system of N particles is in state $z \in \mathcal{A} \subset \mathbb{R}^{3N}$. For the j-th particle $z^{(j)} = \begin{pmatrix} z_{3j-2} & z_{3j-1} & z_{3j} \end{pmatrix}^\top$ we assume that it is in contact with the boundary and that it slides from $z^{(j)} \in \partial \mathcal{S} \subset \mathbb{R}^3$ to $y \in \partial \mathcal{S} \subset \mathbb{R}^3$ instantaneously at time t then the expression

$$\Psi^{(j)}(t, z, y - z^{(j)})$$

presents an approximation for the dissipated energy for $y \approx z^{(j)}$. It is only an approximation since we evaluate the normal force and the roughness only at the initial state z or $z^{(j)}$ of the sliding and not along the path of the sliding from $z^{(j)}$ to y. The precise expression for the dissipated energy will be presented in the energetic problem formulation (S)&(E), see Subsection 2.2.3.

Anyhow we are for the moment interested in the frictional forces only and we announced to model the frictional forces according to the Coulomb friction law. As a motivation we present the formula for a simple one dimensional situation, i.e. $z : [0, T] \to \mathbb{R}$ with constant coefficient of friction μ. The general formula follows in the next paragraph. With the help of the set valued 'Sign'-function, see also Figure 2.2,

$$\mathrm{Sign}(v) := \begin{cases} \{-1\} & \text{for } v < 0 \\ [-1, 1] & \text{for } v = 0 \\ \{+1\} & \text{for } v > 0 \end{cases} \tag{2.2}$$

the Coulomb law is expressed by the inclusion

$$-r \in \sigma(t, z(t))_+ \, \mu \, \mathrm{Sign}(\dot{z}(t)).$$

Here r are the frictional forces that act on the particle. Hence, the Coulomb friction law ensures that the frictional forces are opposite to the sliding direction and proportional to the product of normal force and roughness. If no sliding occurs and the particle sticks then a whole set of forces is possible.

In this paragraph we now present finally our frictional force $\mathcal{D}^{(j)}$ or $\mathcal{D}$, see (2.4) and (2.5) below. In our more general situation we model anisotropic frictional forces by taking the subdifferential of $\Psi^{(j)}$ with respect to the third variable v, i.e. $\partial_v \Psi^{(j)}$.

Figure 2.2: The 'Sign' function

Definition 2.1 (Subdifferential) *Let* $f : \mathbb{R}^n \to (-\infty, \infty]$, *for* $z \in \mathbb{R}^n$ *we define*

$$\partial f(z) := \left\{ v^* \in \mathbb{R}^{n^*} \ : \ v^*(y - z) \leq f(y) - f(z) \ \text{for all } y \in \mathbb{R}^n \right\}.$$

We call the set $\partial f(z) \subset \mathbb{R}^{n^*}$ *the subdifferential of* f *in* z.

From a physical point of view the subdifferential of a convex potential represents all forces which the potential generates. Sometimes it is not possible to derive the exact forces just from the state of the system. That is why we find as a result set valued functions, e.g. if the particle does not slide then many sticking forces are possible but we do not know precisely which is active. To calculate subdifferentials we cite the following useful theorems. We denote by $2^{\mathbb{R}^{3^*}}$ the set of all subsets of the dual space, i.e. $\mathcal{V}^* \in 2^{\mathbb{R}^{3^*}}$ is equivalent to $\mathcal{V}^* \subset \mathbb{R}^{3^*}$.

Theorem 2.2 ([Roc81], Theorem 3F) *Assume* $\mathcal{W}^* \subset \mathbb{R}^{3^*}$ *to be a given compact set and let the function* $f : \mathbb{R}^3 \to \mathbb{R}$ *be defined for* $v \in \mathbb{R}^3$ *via*

$$f(v) := \sup \left\{ w^* v \ : \ w^* \in \mathcal{W}^* \right\}.$$

Then the subdifferential $\partial f \ : \ \mathbb{R}^3 \to 2^{\mathbb{R}^{3^*}}$ *is calculated with the help of the set* $\mathcal{M}^*(v) := \{ w^* \in \mathcal{W}^* \ : \ w^* v = f(v) \}$ *via*

$$\partial f(v) = convex \ hull \ of \ \mathcal{M}^*(v).$$

Theorem 2.3 (Subdifferential $\partial_v \Psi^{(j)}$) *Let the functional* $\Psi^{(j)}$ *be defined as in* (2.1). *Then, writing* m *instead of* $\mathrm{M}^{(j)}(z^{(j)})v$, *the subdifferential* $\partial_v \Psi^{(j)} : [0, T] \times \mathcal{A} \times \mathbb{R}^3 \to 2^{\mathbb{R}^{3^*}}$ *of* Ψ *with respect to the third variable reads*

$$\partial_v \Psi^{(j)}(t, z, v) = \begin{cases} \left\{ \frac{\sigma^{(j)}(t,z)_+}{\|m\|} (m)^\top \mathrm{M}^{(j)}(z^{(j)}) \right\} & \text{if } m \neq 0, z^{(j)} \in \partial \mathcal{S} \\ \left\{ \sigma^{(j)}(t, z)_+ \, v^* \mathrm{M}^{(j)}(z^{(j)}) \ : \ v^* \in \mathcal{B}_1^*(0) \right\} & \text{if } m = 0, z^{(j)} \in \partial \mathcal{S} \\ \{0\} & \text{else.} \end{cases}$$

Remark 2.4 *Note that due to* $\mathrm{M}^{(j)}(z^{(j)})\nu = 0 \in \mathbb{R}^3$ *we find* $w^* \nu = 0$ *for all frictional forces* $w^* \in \partial_v \Psi^{(j)}$. *Thus, the frictional forces do not act in normal direction* $\pm\nu$ *and if they have any effect then only in tangential directions of the boundary, as physically expected.*

Proof: For fixed $z \in \mathcal{A}$ we define the function $f : \mathbb{R}^3 \to \mathbb{R}$ via $f(v) := \|M^{(j)}(z^{(j)})v\|$ and the set

$$\mathcal{W}^* := \left\{ v^* M^{(j)}(z^{(j)}) \in \mathbb{R}^{3^*} \; : \; v^* \in \mathcal{B}_1^*(0) \right\}$$

which obviously is compact. We now check the relation

$$f(v) = \sup \left\{ w^* v \; : \; w^* \in \mathcal{W}^* \right\} \tag{2.3}$$

to be able to apply Theorem 2.2. First it is easy to see by definition $f(v) \geq w^* v$ for arbitrary $v \in \mathbb{R}^3$ and $w^* \in \mathcal{W}^*$. Next we define for $v \in \mathbb{R}^3$ the set

$$\mathcal{M}^*(v) := \begin{cases} \left\{ \frac{1}{\|M^{(j)}(z^{(j)})v\|} \left(M^{(j)}(z^{(j)})v \right)^\top M^{(j)}(z^{(j)}) \right\} & \text{if} \quad M^{(j)}(z^{(j)})v \neq 0 \text{ and} \\ \mathcal{B}_1^*(0) M^{(j)}(z^{(j)}) & \text{if} \quad M^{(j)}(z^{(j)})v = 0. \end{cases}$$

By construction we find $w^* v = f(v)$ for $w^* \in \mathcal{M}^*(v) \subset \mathcal{W}^*$. This proves on the one hand the relation (2.3) and on the other hand, due to Theorem 2.2 and since $\mathcal{M}^*(v)$ is convex, directly the formula $\partial f(v) = \mathcal{M}^*(v)$. Due to $\Psi(t,z,v) = \sigma(t,z) + f(v)$ it is easy to verify the above formula for $\partial_v \Psi^{(j)}$.   ∎

Having the formula for the subdifferential at hand we now finally present the formula for the frictional force for a single particle $\mathcal{D}^{(j)} : [0,T] \times \mathcal{A} \times \mathbb{R}^3 \to 2^{3^*}$ via

$$\mathcal{D}^{(j)}(t,z,v) := -\partial_v \Psi^{(j)}(t,z,v) \subset \mathbb{R}^{3^*}. \tag{2.4}$$

We discuss now some aspects of this expression and make clear that it models an anisotropic Coulomb friction law. If we take for example $M^{(j)} \equiv \begin{pmatrix} \mu & 0 & 0 \\ 0 & \mu & 0 \\ 0 & 0 & 0 \end{pmatrix}$ then we find again the classical isotropic Coulomb friction law. Otherwise let us assume that in the z_1-direction we have twice as much roughness as in the z_2-direction. For this we choose $M_{11}^{(j)} = 2\mu$ and $M_{22}^{(j)} = \mu$ in the above example. In Figure

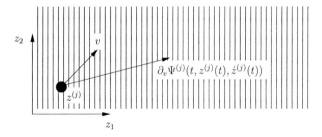

Figure 2.3: Grooved obstacle surface and anisotropic friction

2.3 we indicate such an anisotropic situation by a top view on a grooved obstacle surface. The j-th particle is in contact and slides along the direction $v \in \mathbb{R}^3$ with $v_1 = v_2 = 1$ and $v_3 = 0$. Since $M^{(j)^\top} M^{(j)}$ occurs in the formula for the frictional forces $\partial_v \Psi^{(j)}$ we find a relation of four to one for the corresponding forces, i.e. $\partial_v \Psi^{(j)}(t, z, v) = c(t, z)\begin{pmatrix} 4v_1 & 1v_2 & 0 \end{pmatrix}$ for some function $c \geq 0$ depending on t and z. Up to now we considered the dissipation caused by one particle. The dissipation of the whole system $\Psi : [0, T] \times \mathcal{A} \times \mathbb{R}^{3N} \rightarrow [0, \infty)$ is assumed to be the sum of all single dissipations

$$\Psi(t, z, v) := \sum_{j=1}^{N} \Psi^{(j)}(t, z, v^{(j)}),$$

i.e. the dissipations at the different contact particles do not interact. As a consequence the frictional force $\mathcal{D}(t, z, v) := -\partial_v \Psi(t, z, v) \subset \mathbb{R}^{3N^*}$ satisfies then, for $j = 1, \ldots, N$, the simple formula

$$\left(\mathcal{D}(t, z, v) \right)^{(j)} = \left(\partial_v \Psi(t, z, v) \right)^{(j)} = \partial_{v^{(j)}} \Psi^{(j)}(t, z, v^{(j)}) \quad \subset \mathbb{R}^{3^*}. \tag{2.5}$$

2.1.4 Constraint forces

The forces which prohibit our particle to intrude into the obstacle are called constraint forces, see (2.7). In later chapters the obstacle will have a more general geometry. This motivates us to introduce already here some geometrical terms which are useful to describe the constraint forces also in a more general setting.

Let $\mathcal{A} \subset \mathbb{R}^{3N}$ be a nonempty set. For given $z \in \mathcal{A}$ and $\varepsilon > 0$ we define local tangential cones. That are sets of the form

$$\mathcal{T}_{\mathcal{A}}^{\varepsilon}(z) := \operatorname{clos} \left\{ v \in \mathbb{R}^{3N} \ : \ \exists \lambda > 0, \lambda v + z \in \mathcal{A} \cap \mathcal{B}_{\varepsilon}(z) \right\}.$$

Note that for $0 < \varepsilon_1 < \varepsilon_2$ we have $\mathcal{T}_{\mathcal{A}}^{\varepsilon_1} \subset \mathcal{T}_{\mathcal{A}}^{\varepsilon_2}$. We define the tangential cone of the set $\mathcal{A}$ in z by

$$\mathcal{T}_{\mathcal{A}}(z) := \cap_{\varepsilon > 0} \mathcal{T}_{\mathcal{A}}^{\varepsilon}(z). \tag{2.6}$$

For convex sets $\mathcal{A} \subset \mathbb{R}^{3N}$ the situation is less complicated since for $0 < \varepsilon_1 < \varepsilon_2$ we have $\mathcal{T}_{\mathcal{A}}^{\varepsilon_1}(z) = \mathcal{T}_{\mathcal{A}}^{\varepsilon_2}(z)$ and hence

$$\mathcal{T}_{\mathcal{A}}(z) = \operatorname{clos} \left\{ v \in \mathbb{R}^{3N} \ : \ \exists \lambda > 0, \lambda v + z \in \mathcal{A} \right\}$$
$$= \operatorname{clos} \left\{ \lambda_0(y - z) \in \mathbb{R}^{3N} \ : \ \lambda_0 > 0, y \in \mathcal{A} \right\}.$$

We see the second equality if we take $\lambda = \frac{1}{\lambda_0}$.
After having introduced the tangential cone $\mathcal{T}_{\mathcal{A}}(z) \subset \mathbb{R}^{3N}$ we introduce its dual opponent, the normal cone $\mathcal{N}_{\mathcal{A}}(z) \subset \mathbb{R}^{3N^*}$. This normal cone will also define our constraint forces. We define for $z \in \mathcal{A}$ the set

$$\mathcal{C}(z) := \mathcal{N}_{\mathcal{A}}(z) := \left\{ v^* \in \mathbb{R}^{3N^*} \ : \ v^* v \leq 0 \text{ for all } v \in \mathcal{T}_{\mathcal{A}}(z) \right\}. \tag{2.7}$$

Obviously we have for $z \in \mathrm{int}\mathcal{A}$ the relations $\mathcal{T}_{\mathcal{A}}(z) = \mathbb{R}^{3N}$ and $\mathcal{N}_{\mathcal{A}}(z) = \{0^*\} \subset \mathbb{R}^{3N^*}$. Also the definition of the set of constraint forces simplifies for convex $\mathcal{A}$ to

$$\mathcal{N}_{\mathcal{A}}(z) = \left\{ v^* \in \mathbb{R}^{3N^*} \ : \ v^*(y - z) \leq 0 \text{ for all } y \in \mathcal{A} \right\}. \tag{2.8}$$

Introducing the characteristic function $\mathcal{X}_{\mathcal{A}}(z) := \begin{cases} 0 & \text{for} \quad z \in \mathcal{A}, \\ \infty & \text{for} \quad z \notin \mathcal{A}, \end{cases}$ we deduce for all $z \in \mathcal{A}$ and convex $\mathcal{A}$ the equality

$$\mathcal{N}_{\mathcal{A}}(z) = \partial \mathcal{X}_{\mathcal{A}}(z) \quad \subset \mathbb{R}^{3N^*}. \tag{2.9}$$

In our special situation of a flat boundary, i.e. $\mathcal{S} = \{ y \in \mathbb{R}^3 \ : \ y_3 \geq 0 \}$ and $\mathcal{A} = \left\{ z \in \mathbb{R}^{3N} \ : \ z^{(j)} \in \mathcal{S}, j = 1, \ldots, N \right\}$, we can characterize the set of constraint forces via

$$(\mathcal{N}_{\mathcal{A}}(z))^{(j)} = \mathcal{N}_{\mathcal{S}}(z^{(j)}) \quad \subset \mathbb{R}^{3^*}$$

with $\mathcal{N}_{\mathcal{S}}(z^{(j)}) = 0$ if $z^{(j)} \in \mathrm{int}\mathcal{S}$ and $\mathcal{N}_{\mathcal{S}}(z^{(j)}) = \left\{ \lambda \nu^{\top} \ : \ \lambda \geq 0 \right\}$ if $z^{(j)} \in \partial \mathcal{S}$. Hence, if the j-th particle is in contact then the corresponding normal force has just the same direction as the normal vector.

2.2 Mathematical models

In this section we present the precise mathematical formulation of our problem. Throughout the whole section we will assume $\mathcal{A} \subset \mathbb{R}^{3N}$ to be a non-empty and closed set. For the energy functional we assume $\mathcal{E} \in \mathrm{C}^1([0, T] \times \mathcal{A}, \mathbb{R})$. The dissipation functional $\Psi : [0, T] \times \mathcal{A} \times \mathbb{R}^{3N} \to [0, \infty)$ is assumed to be measurable and bounded on bounded sets. Further, for given $t \in [0, T]$ and $z \in \mathcal{A}$ we assume $\Psi(t, z, \cdot) : \mathbb{R}^{3N} \to [0, \infty)$ to be convex and continuous. Hence, the subdifferential $\partial_v \Psi(t, z, v) \subset \mathbb{R}^{3N^*}$ is non-empty for all $v \in \mathbb{R}^{3N}$.

2.2.1 Differential inclusion

In the previous section we motivated the mathematical formulation as a differential inclusion which corresponds to force balance from a physical point of view.

Problem 2.5 (Differential inclusion) *For given initial time $T_0 \in [0, T)$ and initial state $Z_0 \in \mathcal{A}$ find a positive time span $\Delta \in (0, T - T_0]$ and a solution $z \in \mathrm{W}^{1,\infty}([T_0, T_0 + \Delta], \mathcal{A})$ such that the initial condition $z(T_0) = Z_0$ is satisfied and such that for almost all $t \in [T_0, T_0 + \Delta]$ the following differential inclusion holds*

$$0 \in \mathrm{D}\mathcal{E}(t, z(t)) + \partial_v \Psi(t, z(t), \dot{z}(t)) + \mathcal{N}_{\mathcal{A}}(z(t)) \quad \subset \mathbb{R}^{3N^*}. \tag{DI}$$

In the above formulation we have chosen an initial time T_0 different of 0. The reason is that if for some $0 < T_1 < T_2 \leq T$ the functions z_1 and z_2 satisfy (DI) on the corresponding intervals $[0, T_1]$ and $[T_1, T_2]$ and z_2 further satisfies the initial condition $z_2(T_1) = Z_1$ with $Z_1 := z_1(T_1)$, then the concatenation $z(t) :=$
$$\begin{cases} z_1(t) & \text{for} \quad t \in [0, T_1], \\ z_2(t) & \text{for} \quad t \in [T_1, T_2], \end{cases}$$
satisfies (DI) on the whole interval $[0, T_2]$. Thus, if we assume that a solution exists on the interval $[0, T_0]$, then Problem 2.5 suggests the existence of a local extension of the solution. The same feature holds for all future problem formulations.

The physical motivation for choosing $\mathrm{W}^{1,\infty}$ as function space will be given in Subsection 2.2.3.

2.2.2   Evolutionary variational inequality

The above Problem 2.5 can be equivalently reformulated if we replace the differential inclusion (DI) by a variational inequality. The problem then takes the form of a quasi-variational inequality.

Problem 2.6 (Variational inequality) *For given initial time $T_0 \in [0, T)$ and initial state $Z_0 \in \mathcal{A}$ find a positive time span $\Delta \in (0, T-T_0]$ and a solution $z \in \mathrm{W}^{1,\infty}([T_0, T_0+\Delta], \mathcal{A})$ such that the initial condition $z(T_0) = Z_0$ is satisfied and such that for almost all $t \in [T_0, T_0+\Delta]$ and for all $v \in \mathbb{R}^{3N}$ the following variational inequality holds*

$$0 \leq \mathrm{D}\mathcal{E}(t, z(t))\big(v - \dot{z}(t)\big) + \tilde{\Psi}(t, z(t), v) - \tilde{\Psi}(t, z(t), \dot{z}(t)). \tag{VI}$$

Here we defined $\tilde{\Psi}(t, z, v) := \Psi(t, z, v) + \mathcal{X}_{\mathcal{T}_{\mathcal{A}(z)}}(v)$ and the tangential cone $\mathcal{T}_{\mathcal{A}}(z)$ is defined as in (2.6).

The proof of the equivalence between the Problem 2.5 and 2.6 is postponed, see Theorem 2.9.

2.2.3   Energetic formulation

The following problem formulation was introduced by Mielke and Theil [MiT99, MiT04]. For a general overview and introduction to this formulation see [Mie05]. We give a further selection of articles to illustrate the wide range of applications of this formulation, such as shape-memory alloys [MiP07, AMS08, RTT06], plasticity [MaM08, CHM02, DDM06] and models for fracture [KMZ08, KnM08, DaZ07, DFT05] and delamination [KMR06]. It can be applied to far more general situations and was originally invented to model shape-memory alloys. Even if we present all existence results as solutions of Problem 2.5 with its differential inclusions (DI) we will use the energetic formulation to prove these results.

Problem 2.7 (Energetic problem) *For given initial time $T_0 \in [0,T)$ and initial state $Z_0 \in \mathcal{A}$ find a positive time span $\Delta \in (0, T{-}T_0]$ and a solution $z \in W^{1,\infty}([T_0, T_0{+}\Delta], \mathcal{A})$ such that the initial condition $z(T_0) = Z_0$ is satisfied and such that for all $t \in [T_0, T_0{+}\Delta]$ the following two conditions hold:*

$$\mathcal{E}(t, z(t)) \leq \mathcal{E}(t, y) + \Psi(t, z(t), y - z(t)) \quad \text{for all } y \in \mathcal{A} \tag{S}$$

and $\quad \mathcal{E}(t, z(t)) + \int_{T_0}^{t} \Psi(\tau, z(\tau), \dot{z}(\tau)) \mathrm{d}\tau = \mathcal{E}(T_0, z(T_0)) + \int_{T_0}^{t} \partial_t \mathcal{E}(\tau, z(\tau)) \mathrm{d}\tau.$ (E)

We briefly discuss this formulation. The condition (S) is called the stability condition. It should not be confused with any notation of stability which is linked to the stability with respect to the initial state Z_0. It expresses that the solution should be energetically stable. If we compare the state $z(t)$ of the solution at time t with any other admissible state $y \in \mathcal{A}$ then passing from $z(t)$ to y dissipates at least as much energy as we could possibly gain. Hence it is energetically preferable to remain in the position $z(t)$ at time t. Note that $\Psi(t, z(t), y - z(t))$ is only a rough approximation for the dissipated energy as we already discussed in the physical modeling in Subsection 2.1.3. But for a small distance between z and y the above expression is a quite reasonable approximation for the dissipated energy. This observation motivates the following procedure. Let $z : [0,1] \rightarrow \mathcal{A}$ be the path of the evolving system, then for $\tau_k := k/N, k = 0, \ldots, N$ the energy dissipated by the system could be better approximated by

$$\sum_{k=0}^{N-1} \Psi(\tau_k, z(\tau_k), z(\tau_{k+1}) - z(\tau_k)) = \sum_{k=0}^{N-1} \Psi\left(\tau_k, z(\tau_k), \frac{z(\tau_{k+1}) - z(\tau_k)}{\tau_{k+1} - \tau_k}\right)(\tau_{k+1} - \tau_k).$$

See also Figure 2.4. For the equality we exploited that Ψ is homogenous of de-

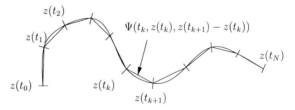

Figure 2.4: Approximation of the dissipated energy

gree one with respect to its third argument, i.e. for $\lambda \geq 0$ we have $\Psi(t, z, \lambda v) = \lambda \Psi(t, z, v)$. Hence, passing to the limit $(N \rightarrow \infty)$ we find the exact expression for the energy dissipated by the system as

$$\int_0^1 \Psi(\tau, z(\tau), \dot{z}(\tau)) \mathrm{d}\tau.$$

Thus, the left hand side of condition (E) describes the energy which we have at time t plus the energy we dissipated during the time interval $[T_0, t]$. While the right hand side expresses the energy we start with at time T_0 plus the work done by external forces which is described by the integral $\int_{T_0}^{t} \partial_t \mathcal{E}(\tau, z(\tau)) d\tau$. Hence, condition (E) is the energy conservation law. Further, having the energetic formulation at hand we can motivate physically our choice of $W^{1,\infty}([0, T], \mathcal{A})$ as function space. Neglecting inertial forces in the mechanical model, see Section 2.1, is equivalent to neglect kinetic energy on the energy level of modeling. Note that in our modeling the energy $\mathcal{E}$ depends only on time t and state z and not on the velocity $\dot{z}$ via the term $m\|\dot{z}(t)\|^2/2$. In fact, we can physically neglect this kinetic term only if the velocity is bounded, i.e. $\|\dot{z}\|_{L^\infty([T_0, T_0+\Delta], \mathbb{R}^{3N})} \leq c$ holds for some constant $1 \gg c > 0$.

The following lemma states the relation between the stability condition (S) and a localized subdifferential formulation. It will be needed in several different chapters.

Lemma 2.8 (Equivalence of (S) **and** (S_{loc})**)** *Let us assume* $\mathcal{A} \subset \mathbb{R}^{3N}$ *to be a convex, closed and non-empty set and* $\mathcal{E} \in C^1([0, T] \times \mathcal{A}, \mathbb{R})$. *For given* $t \in [0, T]$ *and* $z \in \mathcal{A}$ *the function* $\Psi(t, z, \cdot)$ *is assumed to be homogenous of degree one and convex. Then the stability condition*

$$\mathcal{E}(t, z) \leq \mathcal{E}(t, y) + \Psi(t, z, y - z) \quad \text{for all } y \in \mathcal{A} \tag{S}$$

implies the following local stability condition

$$0 \in D\mathcal{E}(t, z) + \partial_v \Psi(t, z, 0) + \mathcal{N}_{\mathcal{A}}(z) \quad \subset \mathbb{R}^{3N^*}. \tag{S_{loc}}$$

If additionally $\mathcal{E}(t, \cdot) : \mathcal{A} \to \mathbb{R}$ *is convex then we have equivalence.*

Proof: To shorten notation we write $\mathcal{E}(z)$ instead of $\mathcal{E}(t, z)$ and also $\Psi(v)$ instead of $\Psi(t, z, v)$. Recalling the characteristic function $\mathcal{X}_{\mathcal{A}}$ from (2.9) we replace the condition (S) equivalently by $(\tilde{S})$:

$$0 \leq \mathcal{E}(y) - \mathcal{E}(z) + \Psi(y - z) + \mathcal{X}_{\mathcal{A}}(y) \quad \text{for all } y \in \mathbb{R}^{3N}. \tag{$\tilde{S}$}$$

We also replace equivalently the local stability condition (S_{loc}) making use of the equality $\mathcal{N}_{\mathcal{A}}(z) = \partial \mathcal{X}_{\mathcal{A}}(z)$ see (2.9):

$$-D\mathcal{E}(z) \in \partial \Psi(0) + \partial \mathcal{X}_{\mathcal{A}}(z) \subset \mathbb{R}^{3N^*}. \tag{$\tilde{S}_{loc}$}$$

Hence, it remains us to prove that $(\tilde{S})$ implies $(\tilde{S}_{loc})$ and that also the opposite holds in the case of convex $\mathcal{E}$.

Let $(\tilde{S})$ hold. For given $v \in \mathbb{R}^{3N}$ and $\lambda \in (0, 1]$ we formally replace the test state $y \in \mathbb{R}^{3N}$ by $\lambda v + z \in \mathbb{R}^{3N}$. After division with λ we find

$$-\frac{1}{\lambda}\Big(\mathcal{E}(\lambda v + z) - \mathcal{E}(z)\Big) \leq \Psi(v) + \frac{1}{\lambda}\mathcal{X}_{\mathcal{A}}(\lambda v + z).$$

Here we used that Ψ is homogenous of degree one, i.e. $\Psi(\lambda v)/\lambda = \Psi(v)$. Since $\mathcal{A}$ is convex $\lambda \mapsto \frac{1}{\lambda}\mathcal{X}_\mathcal{A}(\lambda v + z)$ is monotone decreasing with respect to λ. Thus for $\lambda \leq 1$, we estimate the last term from above by $\mathcal{X}_\mathcal{A}(v + z)$. The right hand-side in the above estimate is then independent of λ and passing to the limit $\lambda \to 0$ on the left-hand side shows

$$-\mathrm{D}\mathcal{E}(z)v \leq \Psi(v) - \Psi(0) + \mathcal{X}_\mathcal{A}(z + v) - \mathcal{X}_\mathcal{A}(z).$$

Here we used $\Psi(0) = \mathcal{X}_\mathcal{A}(z) = 0$. But since $v \in \mathbb{R}^{3N}$ was arbitrary we reformulate this estimate in terms of subdifferentials to find $-\mathrm{D}\mathcal{E}(z) \in \partial\big(\Psi(\cdot) + \mathcal{X}_\mathcal{A}(z + \cdot)\big)(0)$. Using the *Moreau-Rockafellar*-Theorem A.3 for the convex functions $\mathcal{X}_\mathcal{A}$ and Ψ we finally get

$$-\mathrm{D}\mathcal{E}(z) \in \partial\Psi(0) + \partial\mathcal{X}_\mathcal{A}(z) \subset \mathbb{R}^{3N^*}. \tag{$\tilde{\mathrm{S}}_{\mathrm{loc}}$}$$

Thus, we deduced from inequality $(\tilde{\mathrm{S}})$ the inclusion $(\tilde{\mathrm{S}}_{\mathrm{loc}})$. Because of the equivalence of the formulations this shows that (S) implies $(\mathrm{S}_{\mathrm{loc}})$.

It remains us to prove the opposite implication in the special situation of convex $\mathcal{E}$. We start with $(\tilde{\mathrm{S}}_{\mathrm{loc}})$ and replace the derivative $\mathrm{D}\mathcal{E}$ by the subdifferential $\partial\mathcal{E}$ since for convex $\mathcal{E}$ we have $\mathrm{D}\mathcal{E} = \partial\mathcal{E}$. We find $0 \in \partial\mathcal{E}(z) + \partial\Psi(0) + \partial\mathcal{X}_\mathcal{A}(z)$. Again, due to the *Moreau-Rockafellar*-Theorem A.3 we deduce $0 \in \partial\big(\mathcal{E}(\cdot) + \Psi(\cdot - z) + \mathcal{X}_\mathcal{A}(\cdot)\big)(z)$ or, respectively, inequality $(\tilde{\mathrm{S}})$. Thus, we have shown that $(\tilde{\mathrm{S}}_{\mathrm{loc}})$ implies $(\tilde{\mathrm{S}})$ and by the equivalence of the formulations that $(\mathrm{S}_{\mathrm{loc}})$ implies (S). ∎

Theorem 2.9 (Equivalence of problem formulations) *Assume the set $\mathcal{A} \subset \mathbb{R}^{3N}$ to be convex, closed and non-empty. Further let $\mathcal{E} \in \mathrm{C}^1([0,T] \times \mathcal{A}, \mathbb{R})$. We assume the dissipation functional Ψ to be upper semi-continuous and bounded on bounded sets. Further for all $(t,z) \in [T_0, T_0+\Delta] \times \mathcal{A}$ the function $\Psi(t,z,\cdot) : \mathbb{R}^{3N} \to [0,\infty)$ is homogeneous of degree one and satisfies the triangle inequality. Then the following relation between the Problems 2.5, 2.6 and 2.7 holds*

$$(\mathrm{DI}) \Leftrightarrow (\mathrm{VI}) \Leftarrow (\mathrm{S})\&(\mathrm{E}).$$

If further for all $t \in [T_0, T_0+\Delta]$ the function $\mathcal{E}(t,\cdot) : \mathcal{A} \to \mathbb{R}$ is convex, then also the opposite implication holds, i.e. we have equivalence

$$(\mathrm{DI}) \Leftrightarrow (\mathrm{VI}) \Leftrightarrow (\mathrm{S})\&(\mathrm{E})$$

Proof: We start with two preliminary results (2.10) and (2.11) and always assume $z \in \mathrm{W}^{1,\infty}([T_0, T_0+\Delta], \mathcal{A})$. We prove

$$v^*\dot{z}(t) = 0 \tag{2.10}$$

for all $v^* \in \mathcal{N}_\mathcal{A}(z)$ and all $t \in [T_0, T_0+\Delta]$ such that $\dot{z}(t)$ exists. For $n \in \mathbb{N}$ big enough define $v_n := n\big(z(t + \frac{1}{n}) - z(t)\big)$. By our choice of t we have $v_n \to \dot{z}(t)$ for

$n \to \infty$. Further it is easy to check that $v_n \in \mathcal{T}_{\mathcal{A}}(z(t))$ holds if $\mathcal{A}$ is convex. Since the tangential cone is closed we conclude, by the way, the second preliminary result

$$\dot{z}(t) \in \mathcal{T}_{\mathcal{A}}(z(t)). \tag{2.11}$$

Defining $v_n := n\big(z(t - \frac{1}{n}) - z(t)\big) \to -\dot{z}(t)$ and proceeding as above we deduce analogously

$$-\dot{z}(t) \in \mathcal{T}_{\mathcal{A}}(z(t)).$$

Hence for $v^* \in \mathcal{N}_{\mathcal{A}}(z(t))$ we have by definition $v^* \dot{z}(t) \leq 0$ and $v^*(-\dot{z}(t)) \leq 0$ and we deduce the desired equation (2.10).

The first equivalence we prove is between Problem 2.5 and 2.6, i.e. between the differential inclusion (DI) and the variational inequality (VI), which we recall now. We have

$$0 \leq \mathrm{D}\mathcal{E}(t, z(t))\big(v - \dot{z}(t)\big) + \Psi(t, z(t), v) + \mathcal{X}_{\mathcal{T}_{\mathcal{A}(z(t))}}(v) - \Psi(t, z(t), \dot{z}(t)) - \mathcal{X}_{\mathcal{T}_{\mathcal{A}(z(t))}}(\dot{z}(t)) \tag{VI}$$

for almost all $t \in [T_0, T_0 + \Delta]$ and for all $v \in \mathbb{R}^{3N}$. By definition of subdifferentials and by the Moreau-Rockafellar Theorem A.3 the inequality (VI) is equivalent to

$$-\mathrm{D}\mathcal{E}(t, z(t)) \in \partial_v \Psi(t, z(t), \dot{z}(t)) + \partial \mathcal{X}_{\mathcal{T}_{\mathcal{A}(z(t))}}(\dot{z}(t)) \quad \subset \mathbb{R}^{3N^*}.$$

Hence, we establish the equivalence between (VI) and (DI) if we prove the equality $\partial \mathcal{X}_{\mathcal{T}_{\mathcal{A}(z(t))}}(\dot{z}(t)) = \partial \mathcal{X}_{\mathcal{T}_{\mathcal{A}(z(t))}}(z(t)) \stackrel{\mathrm{def}}{=} \mathcal{N}_{\mathcal{A}}(z(t))$. For the first implication we assume $w^* \in \partial \mathcal{X}_{\mathcal{T}_{\mathcal{A}(z(t))}}(\dot{z}(t))$. By definition this is equivalent to have

$$w^*(v - \dot{z}(t)) \leq \mathcal{X}_{\mathcal{T}_{\mathcal{A}(z(t))}}(v) - \mathcal{X}_{\mathcal{T}_{\mathcal{A}(z(t))}}(\dot{z}(t)) = \mathcal{X}_{\mathcal{T}_{\mathcal{A}(z(t))}}(v) \quad \text{for all } v \in \mathbb{R}^{3N}.$$

For the last equality we exploited the second preliminary result $\dot{z}(t) \in \mathcal{T}_{\mathcal{A}}(z(t))$. We show next that $v^* \in \mathcal{N}_{\mathcal{A}}(z(t))$ can be characterized by the same inequality. By definition of the constraint forces $\mathcal{N}_{\mathcal{A}}(z(t))$ we have $v^* v \leq \mathcal{X}_{\mathcal{T}_{\mathcal{A}(z(t))}}(v)$ for all $v \in \mathbb{R}^{3N}$. Further we have shown $v^* \dot{z}(t) = 0$ as preliminary result in (2.10). Thus $v^* \in \mathcal{N}_{\mathcal{A}}(z(t))$ can be characterized also by the same inequality $v^*(v - \dot{z}(t)) \leq \mathcal{X}_{\mathcal{T}_{\mathcal{A}(z(t))}}(v)$ as w^*. This proves $\partial \mathcal{X}_{\mathcal{T}_{\mathcal{A}(z(t))}}(\dot{z}(t)) = \mathcal{N}_{\mathcal{A}}(z(t))$ and thus the equality between (DI) and (VI).

Before we prove the equivalence between (S)&(E) and (DI) we emphasize that the energy equality (E) holds for all $t \in [T_0, T_0 + \Delta]$ if and only if the following local energy equality

$$-\mathrm{D}\mathcal{E}(t, z(t))\dot{z}(t) = \Psi(t, z(t), \dot{z}(t)) \tag{$\mathrm{E}_{\mathrm{loc}}$}$$

holds for almost all $t \in [T_0, T_0 + \Delta]$ (for all $t \in [T_0, T_0 + \Delta]$ for which $\dot{z}(t)$ exists). The equivalence follows from the fundamental theorem of calculus and the chain rule. Let us now assume $z \in \mathrm{W}^{1,\infty}([T_0, T_0 + \Delta], \mathcal{A})$ to be a solution of the (S)&(E) formulation. The aim is to show that z solves the inclusion (DI). In the previous

Lemma 2.8 we have shown that the energetic solution z satisfies (S_{loc}) and above we have shown that z also satisfies (E_{loc}). Combining these two local relations and recalling the preliminary result (2.10) we find that for all $t \in [T_0, T_0+\Delta]$, for which $\dot{z}(t)$ exists, there exists $v^* \in \mathcal{N}_\mathcal{A}(z(t))$ such that

$$(-D\mathcal{E}(t, z(t)) - v^*)(v - \dot{z}(t)) \leq \Psi(t, z(t), v) - \Psi(t, z(t), \dot{z}(t)) \quad \text{for all } v \in \mathbb{R}^{3N}.$$
(2.12)

This inequality is equivalent to the desired differential inclusion (DI) of Problem 2.5.

For the opposite implication we assume $\mathcal{E}(t, \cdot) : \mathcal{A} \to \mathbb{R}$ to be convex for all $t \in [T_0, T_0+\Delta]$ and $z \in W^{1,\infty}([T_0, T_0+\Delta], \mathcal{A})$ to be a solution of Problem 2.5. Thus z satisfies either (DI) or equivalently (2.12) for almost all $t \in [T_0, T_0+\Delta]$. We deduce from (2.12) the local energy equality (E_{loc}) by choosing once $v = 0$ and once $v = 2\dot{z}(t)$ and making use of the homogeneity of degree one of $\Psi(t, z, \cdot)$. Since (E_{loc}) and (E) are equivalent we deduce that z satisfies (E) for almost all times $t \in [T_0, T_0+\Delta]$. We derive the local stability (S_{loc}) from the inequality (2.12) if we choose $v := w + \dot{z}(t)$. The triangle inequality for $\Psi(t, z, \cdot)$ leads us then to

$$\left(-D\mathcal{E}(t, z(t)) - v^*\right)w \leq \Psi(t, z(t), w) \quad \text{for all } w \in \mathbb{R}^{3N}.$$

This is (S_{loc}) for almost all $t \in [T_0, T_0+\Delta]$ and since $\mathcal{E}(t, \cdot)$ is convex we established (S) for almost all $t \in [T_0, T_0+\Delta]$, see Lemma 2.8. Summarizing, the solution z of Problem 2.5 satisfies (S) and (E) for almost all times $t \in [T_0, T_0+\Delta]$. Due to the regularity of $\mathcal{E}$ and due to Ψ being upper semi-continuous and bounded we find the conditions (S) and (E) to hold on the whole interval $[T_0, T_0+\Delta]$. Hence, z solves the energetic Problem 2.7.   ∎

2.3 Stability of the quasi-static path

In this section we show for a rather simple model that quasi-static and dynamic solutions remain close to each other. In the previous sections we have shown that a slow external loading leads to slow solutions of the quasi-static and rate-independent problems. Hence, the kinetic energy that corresponds to these solutions z is quite small and thus the quasi-static solutions z solve a problem which is formally close to the dynamical problem. This is a formal indication that the two solutions remain close to each other and thus that the quasi-static solution is a good approximation of the dynamical solution in the situation of slow external loadings. In this section we now give a rigorous proof of this conjecture, but we restrict ourself to a quite simple situation with one degree of freedom only.

The notion of the 'stability of a quasi-static path' was introduced by a Martins et al. in [MSo*05]. This stability was established in a series of articles for different models in mechanics [PMM07, MMP07, SMR07]. We followed these ideas and especially improved the assumptions on the regularity of the data. The result was already partially published in [SMR07].

2.3.1 Model and definition of stability

We now present our simple model. The admissible set is two-dimensional only and defined via $\mathcal{A} := \{ z \in \mathbb{R}^2 : z_2 \geq 0 \}$. For $z \in \mathcal{A}$ we denote by $z_T := z_2 \in \mathbb{R}$ the tangential component and define the normal component via $z_N := z_2 \in [0, \infty)$.

The evolution of the system is described with respect to the slow loading time t which is linked with respect to the physical time τ via the parameter $\varepsilon > 0$, i.e. $t = \varepsilon \tau$. For example we want to model the behavior of a rail-way bridge crossing a valley. We assume that the characteristic time scale of the bridge (the time scale of the resonance frequencies of the bridge) is in the range of seconds however the external excitation of the bridge acts on a far slower time-scale. We assume that the construction is excited by the movement of the underground which is expressed in terms of meters per century. And since a century consists of $3.153.600.000 = 100 \cdot 365 \cdot 24 \cdot 60 \cdot 60$ seconds we find $\varepsilon = \frac{1}{3.153.600.000} \approx 3 \cdot 10^{-10}$ for the loading rate ε. The question is then which equation describes the evolution of the system with respect to the slow timescale.

For simplicity we assume a linear model for the elasticity. Thus the energy functional is quadratic, $\mathcal{E}(t, z) := \frac{1}{2} z^\top H z - f(t) z$ with a symmetric and positive definite matrix $H \in \mathbb{R}^{2 \times 2}$ and with an external force

$$f \in C^2([0, T], \mathbb{R}^{2^*}). \tag{2.13}$$

Again, we distinguish between normal and tangential components and introduce the notations

$$H = \begin{pmatrix} H_{TT} & H_{TN} \\ H_{NT} & H_{NN} \end{pmatrix}, \quad f(t) = \begin{pmatrix} f_T(t) \\ f_N(t) \end{pmatrix}, \quad r(t) = \begin{pmatrix} r_T(t) \\ r_N(t) \end{pmatrix}. \tag{2.14}$$

Here r_T and r_N represent the reaction forces, i.e. the frictional and constraint forces respectively. With the help of these forces we are able to present the governing equations for the dynamic and quasi-static situation in a common form. The equation of motion in the dynamic case is

$$\varepsilon^2 \ddot{z}(t) + H z(t) - f(t) = r(t), \tag{2.15}$$

where, without loss of generality, we assume a unit mass. The equation of motion in the quasi-static case reads

$$H z(t) - f(t) = r(t). \tag{2.16}$$

We assume an isotropic situation of friction and denote by $\mu > 0$ a constant coefficient of friction. With the help of the set valued function 'Sign' defined in (2.2) we can formulate the Coulomb friction law as follows

$$-r_{\mathsf{T}}(t) \in \mu r_{\mathsf{N}}(t)\mathrm{Sign}\,(\dot{z}_{\mathsf{T}}(t))\,. \tag{2.17}$$

The unilateral contact conditions satisfied by the solutions are given by

$$z_{\mathsf{N}} \geq 0, \quad z_{\mathsf{N}}r_{\mathsf{N}} = 0, \quad r_{\mathsf{N}} \geq 0. \tag{2.18}$$

In the whole section we assume that we are in situations of persistent contact, so that $z_{\mathsf{N}} \equiv 0$. As a consequence we only have one degree of freedom. For example in the dynamic case, the equations (2.15), (2.17) and (2.18) lead to

Problem 2.10 (Dynamic problem with persistent contact) *For $Z_0, V_0 \in \mathbb{R}$ given find a time span $\Delta \in (0,T]$ and a dynamic solution $z_{\mathsf{T}} \in W^{2,\infty}([0,\Delta],\mathbb{R})$ satisfying the initial conditions*

$$z_{\mathsf{T}}(0) = Z_0, \quad \varepsilon \dot{z}_{\mathsf{T}}(0) = V_0, \tag{2.19}$$

and such that, for all $t \in [0,\Delta]$,

$$\varepsilon^2 \ddot{z}_{\mathsf{T}}(t) + H_{\mathsf{TT}}z_{\mathsf{T}}(t) - f_{\mathsf{T}}(t) \in -\mu(H_{\mathsf{NT}}z_{\mathsf{T}}(t) - f_{\mathsf{N}}(t))\mathrm{Sign}\,(\dot{z}_{\mathsf{T}}(t))\,, \tag{2.20}$$

$$H_{\mathsf{NT}}z_{\mathsf{T}}(t) - f_{\mathsf{N}}(t) \geq 0. \tag{2.21}$$

Lemma 2.11 (Existence of a dynamic solution [MMP05]) *Let the function $f \in C^2([0,T],\mathbb{R}^2)$ satisfy*

$$-f_{\mathsf{N}}(0) > H_{\mathsf{NT}}Z_0 \tag{2.22}$$

then there exists $\Delta \in (0,T]$ and a solution $z_{\mathsf{T}} \in W^{2,\infty}([0,\Delta],\mathbb{R})$ of Problem 2.10.

To distinguish the dynamic solution from the quasi-static one, the latter is denoted by $\bar{z}_{\mathsf{T}}$. By taking $\varepsilon = 0$ in (2.20) we formally get the corresponding quasi-static problem with persistent contact.

Problem 2.12 (Quasi-static problem with persistent contact) *For $\bar{Z}_0 \in \mathbb{R}$ find a time span $\Delta \in (0,T]$ and a quasi-static solution $\bar{z}_{\mathsf{T}} \in W^{1,\infty}([0,\Delta],\mathbb{R})$ satisfying*

$$\bar{z}_{\mathsf{T}}(0) = \bar{Z}_0, \tag{2.23}$$

and such that, for all $t \in [0,\Delta]$,

$$H_{\mathsf{TT}}\bar{z}_{\mathsf{T}}(t) - f_{\mathsf{T}}(t) \in -\mu\,(H_{\mathsf{NT}}\bar{z}_{\mathsf{T}}(t) - f_{\mathsf{N}}(t))\,\mathrm{Sign}\,(\dot{\bar{z}}_{\mathsf{T}}(0))\,, \tag{2.24}$$

$$H_{\mathsf{NT}}\bar{z}_{\mathsf{T}}(t) - f_{\mathsf{N}}(t) \geq 0. \tag{2.25}$$

Lemma 2.13 (Existence of a quasi-static solution)
Assume that $\mu > 0$ and $f \in C^1([0,T], \mathbb{R}^2)$. Further our data satisfies

$$|H_{TT}\bar{Z}_0 - f_T(0)| \le \mu \left(H_{NT}\bar{Z}_0 - f_N(0) \right), \tag{2.26}$$

$$-f_N(0) > H_{N,T}\bar{Z}_0 \quad and \tag{2.27}$$

$$H_{TT} - \mu|H_{NT}| > 0. \tag{2.28}$$

Then there exists a positive time span $\Delta \in (0,T]$ and a quasi-static solution $\bar{z}_T \in W^{1,\infty}([0,\Delta], \mathbb{R})$ of the problem (2.23)-(2.25).

Corollary 2.14 *Due to the Assumption (2.27) we can choose $\Delta > 0$ such that there exists $c > 0$ satisfying $c \le H_{NT}\bar{z}(t) - f_N(t)$ for all $t \in [0, \Delta]$.*

Proof: The proof follows directly from Theorem 2.2 in [ScM07], where existence of a quasi-static solution even without the limitation of persistent contact was proven. In the same way the result follows from the second part of existence Theorem 4.2. Note that the Assumptions (O3) and (O5) of Theorem 4.2 correspond to (2.26) and (2.28). The additional property of persistent contact follows from the Assumption (2.27) and the continuity of f and $\bar{z}_T$. ∎

The following notion of stability is not only a stability with respect to the loading rate $\varepsilon > 0$ but also with respect to the initial state.

Definition 2.15 (Stability of a quasi-static path) *Let a quasi-static path, i.e. a solution $\bar{z}_T \in W^{1,\infty}([0,\Delta], \mathbb{R})$ of the quasi-static Problem 2.12 be given. We call the quasi-static path $\bar{z}_T$ stable at $t = 0$, if there exists some positive interval $0 < \Delta_0 \le \Delta$ such that for every $\delta > 0$, we can find constants $C_{ini}(\delta) > 0$ and $C_\varepsilon(\delta) > 0$, such that for each $\varepsilon > 0$ and initial conditions Z_0 and V_0 at $t = 0$ with*

$$|V_0| + |Z_0 - \bar{Z}_0| < C_{ini}(\delta) \quad and \quad \varepsilon < C_\varepsilon(\delta) \tag{2.29}$$

the dynamic solution $z_T \in W^{1,\infty}([0,\Delta], \mathbb{R})$ of Problem 2.10 remains near the quasi-static path in the following sense

$$|\varepsilon\dot{z}_T(t)| + |z_T(t) - \bar{z}_T(t)| < \delta, \tag{2.30}$$

for all $t \in [0, \Delta_0]$.

In the following we assume that f_N always satisfies the Assumptions (2.22) and (2.27), i.e. $-f_N(0) > \max\{H_{N,T}Z_0, H_{N,T}\bar{Z}_0\}$ which guarantees persistent contact on small time intervals. Thus, we focus on the inclusions (2.20) and (2.24) only.

2.3.2 Variation of the derivative of the quasi-static path

In this subsection we prepare a useful technical lemma stating that the velocity of
the quasi-static solution has a bounded variation. On the one hand the proof of
the lemma clarifies in an intuitive way why the solution has a bounded variation.
On the other hand, since it is based on a case study, it is quite technical. There
exist more general regularity results which we could use. But since they are quite
abstract and are based on hysteresis operators we omitted them. We find such a
result in [Kre99] Theorem 8.2. together with [BKS04] Proposition 5.4.
A short calculation shows that the inclusion (2.24) is equivalent to the following
sweeping process formulation

$$-\dot{\bar{z}}_{\mathsf{T}}(t) \in \mathcal{N}_{[g(t),h(t)]}(\bar{z}_{\mathsf{T}}(t)) \subset \mathbb{R}^*, \tag{2.31}$$

where the functions $g, h \in \mathrm{C}^2([0, \Delta], \mathbb{R})$ are defined by

$$g(t) := \frac{f_{\mathsf{T}}(t) + \mu f_{\mathsf{N}}(t)}{H_{\mathsf{TT}} + \mu H_{\mathsf{NT}}} \text{ and } h(t) := \frac{f_{\mathsf{T}}(t) - \mu f_{\mathsf{N}}(t)}{H_{\mathsf{TT}} - \mu H_{\mathsf{NT}}}$$

and $\mathcal{N}_{[g(t),h(t)]}(z) := \{v^* \in \mathbb{R}^* : v^*(z - \tilde{z}) \geq 0 \text{ for all } \tilde{z} \in [g(t), h(t)]\}$ denotes the
corresponding normal cone $z \in [g(t), h(t)]$. Note that due to Corollary 2.14
there exists $\Delta_0, c > 0$ such that $h(t) - g(t) \geq c > 0$ holds for all $t \in [0, \Delta_0]$.
In the following we denote by $\Pi : 0 = \tau_0 < \tau_1 < \cdots < \tau_{N_\Pi} = \Delta_0$, with $N_\Pi \in \mathbb{N}$, a
partition of the interval $[0, \Delta_0]$, and we denote by $\Pi[0, \Delta_0]$ the set of all partitions
of $[0, \Delta_0]$.
For $\Delta_0 \in [0, T]$ the variation of a function of one variable, $f : [0, T] \to \mathbb{R}$, is defined
as

$$\mathrm{var}(f; 0, \Delta_0) := \sup_{\Pi \in \Pi[0, \Delta_0]} \sum_{j=1}^{N_\Pi} |f(\tau_j) - f(\tau_{j-1})|. \tag{2.32}$$

We say that f is of bounded variation on the interval $[0, \Delta_0]$ if $\mathrm{var}(f; 0, \Delta_0) < \infty$
holds. The following lemma is technical and provides us with an auxiliary result
for the proof of the main result, i.e. Theorem 2.17.

Lemma 2.16 (bounded variation of the derivative) *Let the time span $\Delta_0 \in$
$(0, T]$ be given and let the functions $g, h \in \mathrm{C}^1([0, \Delta_0], \mathbb{R})$ satisfy $\mathrm{var}(\dot{g}; 0, \Delta_0) < \infty$
and $\mathrm{var}(\dot{h}; 0, \Delta_0) < \infty$, and also $h(t) - g(t) \geq c > 0$ for some $c > 0$ and all
$t \in [0, \Delta_0]$. Then each solution $z \in \mathrm{W}^{1,\infty}([0, \Delta_0], \mathbb{R})$ of*

$$-\dot{z}(t) \in \mathcal{N}_{[g(t),h(t)]}(z(t)). \tag{2.33}$$

*is differentiable from the right for all $t \in [0, \Delta_0)$ and we define $\dot{z}(t) := \lim_{h \searrow 0} \frac{z(t+h) - z(t)}{h}$.
This right derivative $\dot{z}$ is a right-continuous function with bounded variation,*

$$\mathrm{var}(\dot{z}; 0, \Delta_0) \leq \mathrm{var}(\dot{g}; 0, \Delta_0) + |\dot{g}(0)| + \mathrm{var}(\dot{h}; 0, \Delta_0) + |\dot{h}(0)|. \tag{2.34}$$

Proof: In the first part of the proof we show that the right derivative $\dot{z}(t) = \lim_{h \searrow 0} \frac{z(t+h)-z(t)}{h}$ exists for all $t \in [0, \Delta_0]$ and is a right-continuous function. We fix in the following an arbitrary $t \in [0, \Delta_0)$.

Let us assume first that $g(t) < z(t) < h(t)$. Due to the continuity of g, h and z, there exists $\delta > 0$ such that $g(\tau) < z(\tau) < h(\tau)$ for all $\tau \in [t - \delta, t + \delta]$. By the sweeping process formulation (2.33) we immediately get $\dot{z}(\tau) = 0$ for all $\tau \in [t - \delta, t + \delta]$ and $\dot{z}$ is continuous at t.

Next we examine the situation $z(t) = g(t)$. Due to $h - g \geq c > 0$, there exists a time span $\delta_* > 0$ such that $z(\tau) \neq h(\tau)$ holds for all $\tau \in [t, t + \delta_*]$. After a careful study we see that for $\tau \in [t, t + \delta_*]$ the solution z is monotonically increasing and satisfies

$$z(\tau) = \max_{\tilde{\tau} \in [t,\tau]} g(\tilde{\tau}). \tag{2.35}$$

This typical formula for sweeping processes is the starting point of the rest of the proof, see also Figure 2.5.

Figure 2.5: The function z defined in (2.35)

As a first case we consider $\dot{g}(t) > 0$. Due to the continuity of $\dot{g}$ there exists $0 < \delta \leq \delta_*$ such that $\dot{g} > 0$ holds on $[t, t + \delta)$. Formula (2.35) implies $z \equiv g$ on $[t, t + \delta)$ and consequently the right derivative $\dot{z}$ exists and satisfies $\dot{z}(\tau) = \dot{g}(\tau) > 0$ for all $\tau \in [t, t + \delta)$. In particular it is right-continuous in t.

In the second case of $\dot{g}(t) < 0$ we derive again from the continuity of $\dot{g}$ and (2.35) the existence of some $\delta > 0$ such that $z(\tau) = g(t) = z(t)$ holds on $[t, t+\delta)$. Consequently we have $\dot{z}(\tau) = 0$ for all $\tau \in [t, t + \delta)$.

In the third and last case we assume $\dot{g}(t) = 0$. Then $\dot{z}$ is right-continuous in t and $\dot{z}(t) = 0$ holds. To see this let us fix $\rho > 0$ and choose $\delta \in (0, \delta_*]$ such that $\dot{g}(\tau)_+ := \max\{0, \dot{g}(\tau)\} < \rho$ holds for $\tau \in [t, t + \delta]$. By formula (2.35) we estimate, for $\tau_1, \tau_2 \in [t, t + \delta]$ with $\tau_2 > \tau_1$,

$$0 \leq z(\tau_2) - z(\tau_1) \leq \max_{\tilde{\tau} \in [\tau_1, \tau_2]} g(\tilde{\tau}) - g(\tau_1) \leq \max_{\tilde{\tau} \in [\tau_1, \tau_2]} \int_{\tau_1}^{\tilde{\tau}} \dot{g}(s) \mathrm{d}s \leq \rho(\tau_2 - \tau_1).$$

From this estimate we deduce $|\dot{z}(\tau)| \leq \rho$ for almost all $\tau \in [t, t + \delta]$. Since $\rho > 0$ was arbitrary this proves both, the right-continuity of $\dot{z}$ in t and $\dot{z}(t) = 0$. This completes the situation of $z(t) = g(t)$.

Since $z(t) = h(t)$ is shown analogously, we have now established the existence of the right derivative $\dot{z}(t)$ and its right-continuity for all $t \in [0, T]$. Furthermore we have shown that $\dot{z} > 0$ implies $\dot{z} = \dot{g}$ while $\dot{z} < 0$ implies $\dot{z} = \dot{h}$.

In the second part of the proof we show the estimate (2.34) on the variation of $\dot{z}$. Let us start with the observation that

$$\operatorname{var}(\dot{z}; 0, \Delta_0) = \operatorname{var}(\dot{z}_+; 0, \Delta_0) + \operatorname{var}((-\dot{z})_+; 0, \Delta_0). \tag{2.36}$$

The proof of (2.34) consists now of the estimates

$$\operatorname{var}(\dot{z}_+; 0, \Delta_0) \leq \operatorname{var}(\dot{g}; 0, \Delta_0) + |\dot{g}(0)| \text{ and} \tag{2.37}$$

$$\operatorname{var}((-\dot{z})_+; 0, \Delta_0) \leq \operatorname{var}(\dot{h}; 0, \Delta_0) + |\dot{h}(0)|. \tag{2.38}$$

Since both estimates are shown analogously we content ourselves with the proof of (2.37). Remember the definition of the variation in (2.32). Hence, we assume an arbitrary partition $\Pi : 0 = \tau_0 < \cdots < \tau_{N_\Pi} = \Delta_0$ to be given. We refine this partition in the following. For this we observe first that if $j \in \{1, \ldots, N_\Pi\}$ is such that our solution switches from slip to stick, i.e. $0 < \dot{z}_+(\tau_{j-1}) = \dot{g}(\tau_{j-1})$ and $0 = \dot{z}_+(\tau_j)$ then the left boundary, which pushed the particle, stops at least once meanwhile, i.e. there exists $\tau \in (\tau_{j-1}, \tau_j]$ with $\dot{g}(\tau) = 0$. Otherwise there would exist $\varepsilon > 0$ such that the boundary keeps on moving, i.e. $\dot{g}(\tau) > 0$ for all times $\tau \in [\tau_{j-1}, \tau_j + \varepsilon)$ and due to formula (2.35) this implies constant pushing $\dot{z} \equiv \dot{g} > 0$ on $[\tau_{j-1}, \tau_j + \varepsilon)$ in contradiction to $\dot{z}_+(\tau_j) = 0$. We construct a finer partition Π_f (i.e. $\Pi \subset \Pi_f$). Let $\tau_{j-1}, \tau_j \in \Pi$ and assume that z_+ switches from slipping to sticking, i.e. $\frac{d}{d\tau} z_+(\tau_{j-1}) > 0$ and $\frac{d}{d\tau} z_+(\tau_j) = 0$. Then there exists a time $\tau \in (\tau_{j-1}, \tau_j]$ which satisfies $\dot{g}(\tau) = 0$ and we take $\tau \in \Pi_f$. The resulting new and finer partition is then called Π_f. The index 'f' stands for 'finer'. Our aim is to show that

$$\sum_{j=1}^{N_\Pi} |\dot{z}_+(\tau_j) - \dot{z}_+(\tau_{j-1})| \leq \sum_{k=1}^{N_{\Pi_f}} |\dot{g}(\tau_k) - \dot{g}(\tau_{k-1})| + |\dot{g}(0)| \overset{\text{def}}{\leq} \operatorname{var}(\dot{g}, 0, \Delta_0) + |\dot{g}(0)|. \tag{2.39}$$

We start with $j = N_\Pi$ and apply then one of the four following cases recursively.

Case 1 If $0 = \dot{z}_+(\tau_j) = \dot{z}_+(\tau_{j-1})$ then we obviously have $0 = |\dot{z}_+(\tau_j) - \dot{z}_+(\tau_{j-1})| \leq |\dot{g}(\tau_j) - \dot{g}(\tau_{j-1})|$.

Case 2 If $0 < \dot{z}_+(\tau_j) = \dot{g}(\tau_j)$ and $0 < \dot{z}_+(\tau_{j-1}) = \dot{g}(\tau_{j-1})$ then we have again $|\dot{z}_+(\tau_j) - \dot{z}_+(\tau_{j-1})| \leq |\dot{g}(\tau_j) - \dot{g}(\tau_{j-1})|$.

Case 3 Let us assume $0 = \dot{z}_+(\tau_j)$ and $0 < \dot{z}_+(\tau_{j-1}) = \dot{g}(\tau_{j-1})$. Then, by construction, there exists $\tau \in \Pi_f$ with $\dot{g}(\tau) = 0$. As a consequence we get the estimate $|\dot{z}_+(\tau_j) - \dot{z}_+(\tau_{j-1})| = |\dot{g}(\tau_{j-1})| \leq |\dot{g}(\tau_j) - \dot{g}(\tau)| + |\dot{g}(\tau) - \dot{g}(\tau_{j-1})|$.

Case 4 The most complicated situation is the fourth case of $0 < \dot{z}_+(\tau_j) = \dot{g}(\tau_j)$ and $0 = \dot{z}_+(\tau_{j-1})$. See also Figure 2.6. Thus, in contrast to the first three cases

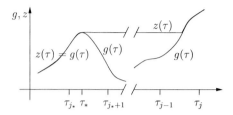

Figure 2.6: The fourth case of our case study

we are here not able to estimate the single difference $|\dot{z}_+(\tau_j) - \dot{z}_+(\tau_{j-1})|$ directly by $|\dot{g}(\tau_j) - \dot{g}(\tau_{j-1})|$ but we need partial sums. Let us assume first that for all $k \leq j-1$ we have $\dot{z}_+(\tau_k) = 0$. Applying the triangle inequality we derive the estimate

$$\sum_{k=1}^{j} |\dot{z}_+(\tau_k) - \dot{z}_+(\tau_{k-1})| = |\dot{z}_+(\tau_j)| = |\dot{g}(\tau_j)| \leq \sum_{k=1}^{j} |\dot{g}(\tau_k) - \dot{g}(\tau_{k-1})| + |\dot{g}(0)|.$$

Otherwise there exists $j_* \in \{0, \dots, j-2\}$ such that $0 < \dot{z}_+(\tau_{j_*}) = \dot{g}(\tau_{j_*})$ and $0 = \dot{z}_+(\tau_k)$ holds for all $j_* < k \leq j-1$. Due to our construction of Π_f there exists $\tau_* \in (\tau_{j_*}, \tau_{j_*+1}]$ with $\dot{g}(\tau) = 0$ and we conclude that $|\dot{g}(\tau_{j_*+1})| + |\dot{g}(\tau_{j_*})| = |\dot{g}(\tau_{j_*+1}) - \dot{g}(\tau_*)| + |\dot{g}(\tau_*) - \dot{g}(\tau_{j_*})|$. Using this we estimate

$$\sum_{k=j_*+1}^{j} |\dot{z}_+(\tau_k) - \dot{z}_+(\tau_{k-1})| = |\dot{z}_+(\tau_j)| + |\dot{z}_+(\tau_{j_*})| = |\dot{g}(\tau_j)| + |\dot{g}(\tau_{j_*})|$$

$$\leq \sum_{k=j_*+2}^{j} |\dot{g}(\tau_k) - \dot{g}(\tau_{k-1})| + |\dot{g}(\tau_{j_*+1}) - \dot{g}(\tau_*)| + |\dot{g}(\tau_*) - \dot{g}(\tau_{j_*})|.$$

Applying recursively one of the four arguments above for decaying index j proves (2.39). Estimating the sum on the right side of (2.39) by $\mathrm{var}(\dot{g}; 0, \Delta_0)$ we conclude

$$\sum_{j=1}^{N_\Pi} |\dot{z}_+(\tau_j) - \dot{z}_+(\tau_{j-1})| \leq \mathrm{var}(\dot{g}; 0, \Delta_0) + |\dot{g}(0)|$$

for an arbitrary partition Π of $[0, \Delta_0]$. Taking the supremum over all partitions establishes formula (2.37). The proof of (2.38) is similar and we omit it. The combination of (2.36)–(2.38) proves the desired estimate (2.34) of the variation of $\dot{z}$. ∎

2.3.3 Stability of the quasi-static path

From Lemma 2.11 and 2.13 we know that there exists a time span $\Delta \in (0, T]$ and solutions $z_\mathsf{T}, \bar{z}_\mathsf{T} : [0, \Delta] \to \mathbb{R}$ of the dynamic problem (2.19)-(2.21) and of the quasi-static problem (2.23)-(2.25), respectively. First we rewrite the inclusions (2.20) and (2.24) by using the functions $\rho, \bar{\rho} : [0, T] \to [-1, 1]$ as follows

$$
\begin{aligned}
\varepsilon^2 \ddot{z}_\mathsf{T}(t) + H_{\mathsf{TT}} z_\mathsf{T}(t) - f_\mathsf{T}(t) &= \rho(t) \mu (H_{\mathsf{TN}} z_\mathsf{T}(t) - f_\mathsf{N}(t)), \\
-\rho(t) &\in \mathrm{Sign}(\dot{z}_\mathsf{T}(t)),
\end{aligned}
\tag{2.40}
$$

$$
\begin{aligned}
H_{\mathsf{TT}} \bar{z}_\mathsf{T}(t) - f_\mathsf{T}(t) &= \bar{\rho}(t) \mu (H_{\mathsf{TN}} \bar{z}_\mathsf{T}(t) - f_\mathsf{N}(t)), \\
-\bar{\rho}(t) &\in \mathrm{Sign}(\dot{\bar{z}}_\mathsf{T}(t)).
\end{aligned}
\tag{2.41}
$$

Note that due to Corollary 2.14 we have for the normal reaction force $\bar{r}_\mathsf{N}(t) = (H_{\mathsf{NT}} \bar{z}_\mathsf{T}(t) - f_\mathsf{N}(t)) \geq c > 0$ for all $t \in [0, \Delta_0]$ and as a consequence $\bar{\rho}(t) = \frac{H_{\mathsf{TT}} \bar{z}_\mathsf{T}(t) - f_\mathsf{T}(t)}{\mu (H_{\mathsf{TN}} \bar{z}_\mathsf{T}(t) - f_\mathsf{N}(t))} \in \mathrm{W}^{1,\infty}([0, \Delta_0], [-1, 1])$. Hence the right derivative of $\bar{\rho}(t)$ which we will denote by $\dot{\bar{\rho}}(t)$ exists for almost all $t \in [0, \Delta_0]$. By differentiating the first line in (2.41) we have

$$
H_{\mathsf{TT}} \dot{\bar{z}}_\mathsf{T} - \dot{f}_\mathsf{T} = \dot{\bar{\rho}} \mu (H_{\mathsf{TN}} \bar{z}_\mathsf{T}(t) - f_\mathsf{N}(t)) + \bar{\rho} \mu (H_{\mathsf{NT}} \dot{\bar{z}}_\mathsf{T} - \dot{f}_\mathsf{N}).
$$

From the right-continuity of the right derivative $\dot{\bar{z}}_\mathsf{T}(t)$ and the inclusion in (2.41) it follows that $\dot{\bar{\rho}}(t) \neq 0$ implies $\dot{\bar{z}}_\mathsf{T}(t) = 0$. This allows us to derive from the above equation the estimate

$$
\frac{|\dot{f}_\mathsf{T}| + \mu |\dot{f}_\mathsf{N}|}{\mu (H_{\mathsf{NT}} \bar{z}_\mathsf{T} - f_\mathsf{N})} \geq |\dot{\bar{\rho}}|.
\tag{2.42}
$$

Hence, due to $\bar{r}_\mathsf{N} = (H_{\mathsf{NT}} \bar{z}_\mathsf{T} - f_\mathsf{N}) \geq c > 0$, see Corollary 2.14, we deduce that $|\dot{\bar{\rho}}|$ is uniformly bounded. Subtracting the equations in (2.40) and (2.41) leads us, for $r_\mathsf{N} := (H_{\mathsf{NT}} z_\mathsf{T} - f_\mathsf{N})$, to

$$
\varepsilon^2 \ddot{z}_\mathsf{T} + H_{\mathsf{TT}}(z_\mathsf{T} - \bar{z}_\mathsf{T}) = \mu \left(\rho r_\mathsf{N} - \bar{\rho} \bar{r}_\mathsf{N} \right) = \mu (\rho - \bar{\rho}) r_\mathsf{N} + \mu \bar{\rho} (r_\mathsf{N} - \bar{r}_\mathsf{N}).
\tag{2.43}
$$

Before multiplying the above equation by $(\dot{z}_\mathsf{T} - \dot{\bar{z}}_\mathsf{T})$, we observe that the inclusion in (2.40) is equivalent to $-\dot{z}_\mathsf{T}(y - \rho) \leq 0$, for all $y \in [-1, 1]$, and (2.41) is equivalent to $-\dot{\bar{z}}_\mathsf{T}(\bar{y} - \bar{\rho}) \leq 0$, for all $\bar{y} \in [-1, 1]$. Choosing $y = \bar{\rho}$ and $\bar{y} = \rho$ we get the monotonicity condition

$$
(\dot{z}_\mathsf{T} - \dot{\bar{z}}_\mathsf{T})(\rho - \bar{\rho}) \leq 0.
$$

Multiplying then (2.43) by $(\dot{z}_\mathsf{T} - \dot{\bar{z}}_\mathsf{T})$, we are immediately led to the estimate

$$
\varepsilon^2 \ddot{z}_\mathsf{T}(\dot{z}_\mathsf{T} - \dot{\bar{z}}_\mathsf{T}) + H_{\mathsf{TT}}(z_\mathsf{T} - \bar{z}_\mathsf{T})(\dot{z}_\mathsf{T} - \dot{\bar{z}}_\mathsf{T}) \leq \mu \bar{\rho}(r_\mathsf{N} - \bar{r}_\mathsf{N})(\dot{z}_\mathsf{T} - \dot{\bar{z}}_\mathsf{T}).
\tag{2.44}
$$

This estimate is rewritten after some rearrangements as

$$
\frac{\mathrm{d}}{\mathrm{d}t} \left(\frac{\varepsilon^2}{2} (\dot{z}_\mathsf{T})^2 \right) + \frac{\mathrm{d}}{\mathrm{d}t} \frac{(H_{\mathsf{TT}} - \mu \bar{\rho} H_{\mathsf{NT}})(z_\mathsf{T} - \bar{z}_\mathsf{T})^2}{2} + \frac{\mu H_{\mathsf{NT}}(z_\mathsf{T} - \bar{z}_\mathsf{T})^2}{2} \frac{\mathrm{d}}{\mathrm{d}t} \bar{\rho} \leq \varepsilon^2 \ddot{z}_\mathsf{T} \dot{\bar{z}}_\mathsf{T}
\tag{2.45}
$$

Next we integrate the above estimate and further use the estimate (2.42) on $\dot{\rho}$ to obtain

$$\frac{\varepsilon^2}{2}(\dot{z}_\mathsf{T}(t))^2 + \frac{H_{\mathsf{TT}} - \mu|H_{\mathsf{NT}}|}{2}(z_\mathsf{T}(t) - \bar{z}_\mathsf{T}(t))^2 \leq C \int_0^t (z_\mathsf{T}(s) - \bar{z}_\mathsf{T}(s))^2 \, ds$$

$$+ \varepsilon^2 \int_0^t \ddot{z}_\mathsf{T}(s)\dot{z}_\mathsf{T}(s) \, ds \qquad (2.46)$$

$$+ \frac{\varepsilon^2}{2}V_0^2 + \frac{H_{\mathsf{TT}} + \mu|H_{\mathsf{NT}}|}{2}(Z_0 - \bar{Z}_0)^2$$

$$(2.47)$$

for some finite constant $C > 0$ depending on $\dot{\rho}$. Next we apply Gronwall's Lemma to estimate $(z_\mathsf{T}(t) - \bar{z}_\mathsf{T}(t))^2$. In a first step we divide both sides by $\frac{H_{\mathsf{TT}} - \mu|H_{\mathsf{NT}}|}{2}$, which is positive due to the assumption (2.28). We omit the exact and lengthy result and we represent it in the following simplified form. There exist positive constants $C_1 < \infty$ and $C_2 < \infty$, depending on the data H, μ and f only such that

$$(z_\mathsf{T}(t) - \bar{z}_\mathsf{T}(t))^2 \leq C_1 G(\varepsilon, t) \exp(C_2 t)$$

holds, with $G(\varepsilon, t) := \varepsilon^2 \left| \int_0^t \ddot{z}_\mathsf{T}(s)\dot{z}_\mathsf{T}(s) \, ds \right| + \frac{\varepsilon^2}{2}V_0^2 + (Z_0 - \bar{Z}_0)^2$. We can use this to estimate the first integral on the right hand side of (2.47) as follows, $\int_0^t (z_\mathsf{T}(s) - \bar{z}_\mathsf{T}(s))^2 \, ds \leq C_3 G(\varepsilon, t)$ with C_3 depending on H, μ, f and Δ_0 only. By this estimate there exists a constant $C_4(H, \mu, f, \Delta_0)$ such that

$$\frac{\varepsilon^2}{2}(\dot{z}_\mathsf{T}(t))^2 + \frac{H_{\mathsf{TT}} - \mu|H_{\mathsf{NT}}|}{2}(z_\mathsf{T}(t) - \bar{z}_\mathsf{T}(t))^2$$

$$\leq C_4 \left(\varepsilon^2 \left| \int_0^t \ddot{z}_\mathsf{T}(s)\dot{z}_\mathsf{T}(s) \, ds \right| + \frac{\varepsilon^2}{2}V_0^2 + (Z_0 - \bar{Z}_0)^2 \right) \qquad (2.48)$$

holds for all $t \in [0, \Delta_0]$. The remaining task is to estimate the integral in (2.48). We have already proved in Lemma 2.16 that the right derivate $\dot{z}_\mathsf{T}$ is right-continuous and has bounded variation, hence we can transform the differential measure of $\dot{z}_\mathsf{T}\dot{z}_\mathsf{T}$ as follows

$$d(\dot{z}_\mathsf{T}\dot{z}_\mathsf{T}) = \dot{z}_\mathsf{T} d\dot{z}_\mathsf{T} + \dot{z}_\mathsf{T} d\dot{z}_\mathsf{T} = \dot{z}_\mathsf{T} d\dot{z}_\mathsf{T} + \dot{z}_\mathsf{T}\dot{z}_\mathsf{T} dt,$$

see also the monograph [Mon93]. Using this formula we can integrate by parts and estimate as follows

$$\varepsilon^2 \int_0^t \ddot{z}_\mathsf{T}(s)\dot{z}_\mathsf{T}(s) \, ds = -\varepsilon^2 \int_0^t \dot{z}_\mathsf{T}(s) d\dot{z}_\mathsf{T}(s) + \varepsilon^2 \dot{z}_\mathsf{T}(t)\bar{z}_\mathsf{T}(t) - \varepsilon^2 V_0 \dot{z}_\mathsf{T}(0)$$

$$\leq \varepsilon^2 \|\dot{z}_\mathsf{T}\|_{\mathrm{L}^\infty(0,\Delta_0)} \mathrm{var}(\dot{z}_\mathsf{T}, 0, t) + \frac{\varepsilon^2}{4C_4}(\dot{z}_\mathsf{T}(t))^2 + \varepsilon^2 C_4 (\dot{z}_\mathsf{T}(t))^2$$

$$+ \frac{\varepsilon^2}{2}V_0^2 + \frac{\varepsilon^2}{2}|\dot{z}_\mathsf{T}(0)|^2.$$

Before introducing this estimate in (2.48) we recall that the quasi-static solution $\bar{z}_T$ is Lipschitz-continuous and that $\mathrm{var}(\dot{\bar{z}}_T, 0, t) \leq \mathrm{var}(\dot{\bar{z}}_T, 0, \Delta_0) < \infty$. Hence there exists a positive constant C_5 such that

$$\frac{\varepsilon^2}{4}(\dot{z}_T(t))^2 + \frac{H_{TT} - \mu|H_{NT}|}{2}(z_T(t) - \bar{z}_T(t))^2 \leq \frac{\varepsilon^2}{8}\|\dot{z}_T\|^2_{L^\infty(0,\Delta)} + \varepsilon^2 C_5$$
$$+ C_4\left(\varepsilon^2 V_0^2 + (Z_0 - \bar{Z}_0)^2\right). \quad (2.49)$$

Since the above inequality holds for all times $t \in [0, \Delta_0]$ it remains correct if we take the L^∞-norm on the left-hand side. Then we find $\frac{\varepsilon^2}{8}\|\dot{z}_T\|^2_{L^\infty(0,\Delta_0)} \leq \varepsilon^2 C_5 + C_4\left(\frac{\varepsilon^2}{2}V_0^2 + (Z_0 - \bar{Z}_0)^2\right)$. Introducing this estimate again into (2.49) and multiplying the inequality with $\frac{1}{2}$ we derive the final estimate

$$\frac{\varepsilon^2}{8}(\dot{z}_T(t))^2 + \frac{H_{TT} - \mu|H_{NT}|}{4}(z_T - \bar{z}_T)^2(t) \leq \varepsilon^2 C_5 + C_4\left(\frac{\varepsilon^2}{2}V_0^2 + (Z_0 - \bar{Z}_0)^2\right)$$
$$\text{for all } t \in [0, \Delta_0].$$
$$(2.50)$$

Note that the estimate (2.50) implies the stability of the quasi-static path as defined in 2.15. We now summarize our results in

Theorem 2.17 (Stability of the Quasi-Static Path) *Let the stiffness matrix* H *be positive definite and the coefficient of friction* $\mu > 0$ *be such that* $H_{TT} > \mu|H_{NT}|$ *holds. In addition, the initial state* $\bar{Z}_0$ *satisfies* $|H_{TT}\bar{Z}_0 - f_T(0)| \leq \mu(H_{NT}\bar{Z}_0 - f_N(0))_+$, *and the external force* f *satisfies* $f \in C^2([0,T], \mathbb{R}^2)$, *var*$(\dot{f}; 0, t) < \infty$ *and* $-f_N(0) > \max\{H_{N,T}\bar{Z}_0, H_{N,T}Z_0\}$. *Then the dynamic and the quasi-static Problems 2.10 and 2.12 with persistent contact have solutions and the quasi-static path is stable at* $t = 0$ *in the sense of Definition 2.15.*

Chapter 3

Varying friction - basic mathematical strategy

stone   ice   stone   stone   ice   stone

Figure 3.1: Varying frictional force: varying convexity of dissipation Ψ

3.1 Overview

In this chapter we deal with the problem of a varying frictional force or mathematically with the problem of a dissipation functional Ψ whose convexity depends on the state z and time t. At the same time this chapter is the fundament for following chapters. We introduce the basic mathematical strategy which in its characteristic features will not be changed. For this reason the whole presentation and notation is quite generic and divided into many small results such that we can reuse them in later chapters.

From a modeling point of view we establish a first existence result for a simple friction problem. The problem consists of a convex admissible set $\mathcal{A} \subset \mathbb{R}^{3N}$ and it is assumed that all particles do not switch between contact and non-contact. So either they are out of contact and remain out of contact or they are constrained to the boundary. As a consequence we can assume the dissipation functional to be continuous with respect to the state z.

The basic method consists of three steps:

    1. construction of approximative solutions,

2. convergence of approximative solutions,

3. proving that the limit function is a solution.

Whenever we will prove an existence result in this work, the first and third step and big parts of the second will always be the same. Due to their generic presentation in this section we will not have to repeat them but just refer to these generic versions. In the sections 3.4 *Approximative solutions*, 3.5 *Convergence* and 3.6 *Existence* we present a method to construct solutions of the energetic problem formulated in section 3.2. Even if the three sections 3.4–3.6 are built on each other we kept them self-contained and repeat at the beginning of each section the results of the previous one in form of assumptions.

3.2 Energetic formulation of the problem

In this section we formulate the energetic problem we want to solve. The formulation is kept general or generic enough such that all models that we will be interested in are included.

Our problem is time-dependent and we fix some initial time $\hat{T} > 0$ and some time span $\hat{\Delta} > 0$. The elastic system we are interested in will be described by some $z \in \mathbb{R}^{3N}$. Further we introduce a convex set $\hat{A} \subset \mathbb{R}^{3N}$ which describes the set of all admissible states of the system. A solution will be a function $z : [\hat{T}, \hat{T} + \hat{\Delta}] \to \hat{A}$. The energetic formulation is based on a *energy functional* $\hat{\mathcal{E}}(t, z)$ and a *dissipation functional* $\hat{\Psi}(t, z, \dot{z})$. (We denote by $\dot{z}$ the time-derivative, i.e. $\dot{z} = \dfrac{\mathrm{d}}{\mathrm{d}t}z$.) The functional $\hat{\mathcal{E}}$ describes the energy that is stored in the system at time t if the particle is in the position $z \in \hat{A}$. The dissipation functional $\hat{\Psi}(t, z, v)$ describes how much energy the systems dissipates if it changes its state at time t from z to $z + \delta v$ for $|\delta| \ll 1$.

Problem 3.1 (Energetic problem) *For a given initial value $\hat{z}_0 \in \hat{A}$ and time interval $[\hat{T}, \hat{T} + \hat{\Delta}]$ find a solution $z \in \mathrm{W}^{1,\infty}([\hat{T}, \hat{T} + \hat{\Delta}], \hat{A})$ such that $z(\hat{T}) = \hat{z}_0$ and for all $t \in [\hat{T}, \hat{T} + \hat{\Delta}]$ the following two conditions hold:*

$$\hat{\mathcal{E}}(t, z(t)) \leq \hat{\mathcal{E}}(t, y) + \hat{\Psi}(t, z(t), y - z(t)) \qquad \text{for all } y \in \hat{A} \qquad \text{(S)}$$

$$\text{and} \qquad \hat{\mathcal{E}}(t, z(t)) + \int_{\hat{T}}^{t} \hat{\Psi}(\tau, z(\tau), \dot{z}(\tau)) \mathrm{d}\tau = \hat{\mathcal{E}}(\hat{T}, z(\hat{T})) + \int_{\hat{T}}^{t} \partial_t \hat{\mathcal{E}}(\tau, z(\tau)) \mathrm{d}\tau. \quad \text{(E)}$$

3.3 Assumptions and existence result

We now present all assumptions that we need to derive a first existence result. Since we present the problem here in a quite general way and use a simplified proof we will need quite many assumptions. Later on, when we consider our concrete problem, several assumptions follow directly from the modeling or will not be needed due to an improved method. Let $m, n \in \mathbb{N}$ and the time interval $[\hat{T}, \hat{T} + \hat{\Delta}]$ be given. We assume that the admissible set

$$\hat{\mathcal{A}} \subset \mathbb{R}^{3N} \text{ is closed and convex.} \tag{3.1}$$

Further we assume that the energy functional $\hat{\mathcal{E}} : [\hat{T}, \hat{T} + \hat{\Delta}] \times \hat{\mathcal{A}} \to [0, \infty)$ satisfies

$$\hat{\mathcal{E}} \in C^2 \left([\hat{T}, \hat{T} + \hat{\Delta}] \times \hat{\mathcal{A}}, [0, \infty) \right). \tag{3.2}$$

We denote by $\hat{H}(t, z) \in \mathbb{R}^{3N \times 3N}$ the Hessian matrix of $\hat{\mathcal{E}}$,i.e. $\hat{H}(t, z) := \frac{d^2}{dz^2}\hat{\mathcal{E}}(t, z)$ and now assume that $\hat{\mathcal{E}}$ is $\hat{\alpha}$-uniformly convex in its second variable, i.e., there exists a positive constant $\hat{\alpha}_* > 0$ such that $\hat{\alpha}(t, z) := \min \left\{ v^\top \hat{H}(t, z)v \ : \ v \in \mathbb{S}^{3N-1} \right\}$ satisfies

$$\hat{\alpha}(t, z) \geq \hat{\alpha}_* \qquad \text{for all } (t, z) \in [\hat{T}, \hat{T} + \hat{\Delta}] \times \hat{\mathcal{A}}. \tag{3.3}$$

Here $\mathbb{S}^{n-1} := \left\{ v \in \mathbb{R}^{3N} \ : \ \|v\| = 1 \right\}$ defines the unit sphere in $\mathbb{R}^{3N}$. We assume that for all $z \in \hat{\mathcal{A}}$ the dissipation functional $\hat{\Psi} : [\hat{T}, \hat{T} + \hat{\Delta}] \times \hat{\mathcal{A}} \times \mathbb{R}^{3N} \to [0, \infty)$ satisfies

$$
\begin{aligned}
\hat{\Psi}(\cdot, z, \cdot) &\in C \left([\hat{T}, \hat{T} + \hat{\Delta}] \times \mathbb{R}^{3N}, [0, \infty) \right) \qquad \text{(weak version)}, \\
\hat{\Psi} &\in C \left([\hat{T}, \hat{T} + \hat{\Delta}] \times \hat{\mathcal{A}} \times \mathbb{R}^{3N}, [0, \infty) \right) \qquad \text{(strong version)}.
\end{aligned}
\tag{3.4}
$$

If we do not give any explicit comment then we assume (3.4) in its weak version only. Further we require that for all $t \in [\hat{T}, \hat{T} + \hat{\Delta}]$ and $x \in \hat{\mathcal{A}}$ the function $\hat{\Psi}(t, x, \cdot) : \mathbb{R}^{3N} \to [0, \infty)$ is positively homogeneous of degree one, i.e.

$$\hat{\Psi}(t, x, \lambda v) = \lambda \hat{\Psi}(t, x, v) \qquad \text{for all } \lambda \geq 0, v \in \mathbb{R}^{3N} \tag{3.5}$$

and satisfies the triangle inequality, i.e.

$$\hat{\Psi}(t, x, v_1 + v_2) \leq \hat{\Psi}(t, x, v_1) + \hat{\Psi}(t, x, v_2) \qquad \text{for all } v_1, v_2 \in \mathbb{R}^{3N}. \tag{3.6}$$

We assume global Lipschitz bounds as follows: there exist constants $\hat{c}_1, \hat{c}_2, \hat{q} > 0$ such that

$$\frac{1}{2}\|\partial_t D\hat{\mathcal{E}}(\tau, z)\| \leq \hat{c}_1 \text{ for all } \tau \in [\hat{T}, \hat{T} + \hat{\Delta}], z \in \hat{\mathcal{A}}, \tag{3.7}$$

$$\left| \hat{\Psi}(t_2, z, v) - \hat{\Psi}(t_1, z, v) \right| \leq \hat{c}_2 |t_2 - t_1| \text{ for all } z \in \hat{\mathcal{A}}, v \in \mathbb{S}^{3N-1}, t_1, t_2 \in [\hat{T}, \hat{T} + \hat{\Delta}], \tag{3.8}$$

$$\left| \hat{\Psi}(t, z_2, v) - \hat{\Psi}(t, z_1, v) \right| \leq \hat{q}\|z_2 - z_1\| \text{ for all } z_{1,2} \in \hat{\mathcal{A}}, v \in \mathbb{S}^{3N-1}, t \in [\hat{T}, \hat{T} + \hat{\Delta}]. \tag{3.9}$$

The previous Assumption (3.9) is a clear simplification, since in our modeling the dissipation is not continuous with respect to the state z due to the passage from contact to non-contact. The crucial assumption for the constants $\hat{\alpha}_*, \hat{q} > 0$ defined in (3.3) and (3.9) is the following:

$$\hat{\alpha}_* > \hat{q}. \tag{3.10}$$

Finally we assume for the initial values $\hat{T}$ and $\hat{z}_0 \in \hat{\mathcal{A}}$ stability, i.e.

$$\hat{\mathcal{E}}(\hat{T}, \hat{z}_0) \leq \hat{\mathcal{E}}(\hat{T}, y) + \hat{\Psi}(\hat{T}, \hat{z}_0, y - \hat{z}_0) \quad \text{for all } y \in \hat{\mathcal{A}}. \tag{3.11}$$

Theorem 3.2 (Existence of a solution) *We assume* (3.1)–(3.11) *with the Assumption* (3.4) *in its strong version, then there exists a solution* $z \in W^{1,\infty}([\hat{T}, \hat{T} + \hat{\Delta}], \hat{\mathcal{A}})$ *of the Energetic Problem 3.1. Further, we have for almost all* $t \in [\hat{T}, \hat{T} + \hat{\Delta}]$ *the estimate*

$$\|\dot{z}(t)\| \leq \frac{\hat{c}_1 + \hat{c}_2}{\hat{\alpha}_* - \hat{q}}. \tag{3.12}$$

Proof: See Corollary 3.27. ∎

3.4 Approximative solutions

In this section we construct approximative solutions and prove their uniqueness. The sections relies on the Assumptions (3.1)– (3.6) only and its final result is Theorem ??.

Lemma 3.3 (Convexity of $\hat{\Psi}$) *Due to the homogeneity* (3.5) *and the triangle inequality* (3.6) *the dissipation functional* $\hat{\Psi}$ *is convex with respect to its third variable, i.e. for all* $t \in [\hat{T}, \hat{T} + \hat{\Delta}], x \in \hat{\mathcal{A}}, \lambda \in [0,1]$ *and* $v_1, v_2 \in \mathbb{R}^{3N}$ *we find*

$$\hat{\Psi}(t, x, \lambda v_1 + (1-\lambda)v_2) \leq \lambda \hat{\Psi}(t, x, v_1) + (1-\lambda)\hat{\Psi}(t, x, v_2). \tag{3.13}$$

Proof: From the triangle inequality (3.6) we deduce

$$\hat{\Psi}(t, x, \lambda v_1 + (1-\lambda)v_2) \leq \hat{\Psi}(t, x, \lambda v_1) + \hat{\Psi}(t, x, (1-\lambda)v_2).$$

Inequality (3.13) now follows directly from the Assumption (3.5). ∎

Definition 3.4 (α-uniform convexity) *Let* $\alpha > 0$ *be given. We call a function* $f : \mathbb{R}^{3N} \to \mathbb{R}$ α-*uniformly convex if for all* $x_1, x_2 \in \mathbb{R}^{3N}$ *and all* $\lambda \in [0,1]$ *we have*

$$\lambda f(x_2) + (1-\lambda)f(x_1) - \frac{\alpha}{2}\lambda(1-\lambda)\|x_1 - x_2\|^2 \geq f(x_1 + \lambda(x_2 - x_1)). \tag{3.14}$$

Lemma 3.5 (α-uniform convexity) *Let $f \in C^2\left(\mathbb{R}^{3N}, \mathbb{R}\right)$ and $\alpha > 0$ be given. Then f being α-uniformly convex is equivalent to the following convexity assumption:*

$$v^\top D^2 f(x)v \geq \alpha \|v\|^2 \tag{3.15}$$

holds for all $x, v \in \mathbb{R}^{3N}$.

Proof: We start with (3.15) and derive the α-uniform convexity of f. For arbitrary $a, b \in \mathbb{R}^{3N}$ we define $x : [0, 1] \to \mathbb{R}^{3N}$ by $x(s) := a + s(b - a)$. With the help of this notation and applying the fundamental theorem of calculus twice we now deduce

$$f(b) - f(a) = \int_0^1 Df(x(s))(b-a)\,ds$$
$$= \int_0^1 \int_0^s (b-a)^\top D^2 f(x(r))(b-a)\,dr\,ds + Df(a)(b-a).$$

We use the above convexity assumption to estimate the integrand from below by $\alpha\|b - a\|^2$. After integration this leads us to

$$f(b) - f(a) \geq \frac{\alpha}{2}\|b - a\|^2 + Df(a)(b - a). \tag{3.16}$$

We introduce the notation $x_\lambda := x_1 + \lambda(x_2 - x_1)$. The above α-uniform convexity condition (3.14) is then equivalent to

$$\lambda\left(f(x_2) - f(x_\lambda)\right) + (1 - \lambda)\left(f(x_1) - f(x_\lambda)\right) \geq \lambda(1 - \lambda)\frac{\alpha}{2}\|x_2 - x_1\|^2$$

for all $\lambda \in [0, 1], x_1, x_2 \in \mathbb{R}^{3N}$. We estimate the two differences on the left side with the help of our intermediate result (3.16) by

$$\lambda\left(f(x_2) - f(x_\lambda)\right) \geq \lambda(1 - \lambda)^2\frac{\alpha}{2}\|x_2 - x_1\|^2 - \lambda(1 - \lambda)Df(x_\lambda)(x_2 - x_1)$$

and

$$(1 - \lambda)\left(f(x_1) - f(x_\lambda)\right) \geq (1 - \lambda)\lambda^2\frac{\alpha}{2}\|x_2 - x_1\|^2 - \lambda(1 - \lambda)Df(x_\lambda)(x_2 - x_1).$$

Adding these estimates leads us exactly to the above equivalent α-uniform convexity condition.

Let us now assume that (3.15) does not hold for some $x, v \in \mathbb{R}^{3N}$. Due to the continuity of $D^2 f$ there exist $r > 0$ and $\varepsilon > 0$ such that $v^\top D^2 f(y)v \leq (\alpha - \varepsilon)\|v\|^2$ holds for all $y \in \mathcal{B}_r(x)$. We continue as in the first part of the proof to derive a contradiction to the definition of α-uniform convexity, i.e. there exist $x_2, x_1 \in \mathbb{R}^{3N}$ and $\varepsilon > 0$ such that for all $\lambda \in [0, 1]$ we have

$$\lambda f(x_2) + (1 - \lambda)f(x_1) - f(x_\lambda) \leq \frac{\alpha - \varepsilon}{2}\lambda(1 - \lambda)\|x_1 - x_2\|^2.$$

This contradicts (3.14) ∎

Corollary 3.6 (α-uniform convexity of $\hat{\mathcal{E}}$) *Due to our uniform convexity condition* (3.3) *on $\hat{\mathcal{E}}$ and the above lemma we find that the function $\hat{\mathcal{E}}(t,\cdot) : \hat{\mathcal{A}} \to \mathbb{R}$ is α-uniformly convex on $\hat{\mathcal{A}}$ for all $t \in [\hat{T}, \hat{T} + \hat{\Delta}]$, i.e.*

$$\hat{\mathcal{E}}(t, z_1 + \lambda(z_2 - z_1)) \leq \lambda \hat{\mathcal{E}}(t, z_2) + (1 - \lambda)\hat{\mathcal{E}}(t, z_1) - \lambda(1 - \lambda)\frac{\hat{\alpha}_*}{2}\|z_2 - z_1\|^2$$

holds for all $z_1, z_2 \in \hat{\mathcal{A}}$ and $\lambda \in [0, 1]$ with $\hat{\alpha}_* > 0$ being defined as in Assumption (3.3).

Lemma 3.7 (Coercivity of $\hat{\mathcal{E}}$) *Let the regularity and the convexity Assumptions* (3.2) *and* (3.3) *on $\hat{\mathcal{E}}$ and the Assumption* (3.1) *on the set $\hat{\mathcal{A}}$ hold. Then the energy functional $\hat{\mathcal{E}}(t,\cdot) : \hat{\mathcal{A}} \to \mathbb{R}$ is coercive for each $t \in [\hat{T}, \hat{T} + \hat{\Delta}]$, i.e.*

$$\hat{\mathcal{E}}(t, z) \to \infty \quad \text{for} \quad \|z\| \to \infty.$$

Remark 3.8 *The Assumption* (3.3) *about the α-uniform convexity of $\hat{\mathcal{E}}(t,\cdot)$ is sufficient to prove coercivity, since* (3.3) *implies for each $t \in [\hat{T}, \hat{T} + \hat{\Delta}]$ continuity of $\hat{\mathcal{E}}(t,\cdot) : \hat{\mathcal{A}} \to \mathbb{R}$. But we will not prove this implication and just use Assumption* (3.2) *for simplicity.*

Proof: Let us fix $t \in [\hat{T}, \hat{T} + \hat{\Delta}]$ and an arbitrary $\hat{z}_0 \in \hat{\mathcal{A}}$. For each sequence $(z_n)_{n \in \mathbb{N}} \in \hat{\mathcal{A}}$ with $\|z_n\| \to \infty$ there exists $n_0 \in \mathbb{N}$ such that $\rho_n := \|z_n - \hat{z}_0\| > 1$ for all $n \geq n_0$. Due to the convexity of $\hat{\mathcal{A}}$ we have $x_n := \hat{z}_0 + \lambda_n(z_n - \hat{z}_0) \in \hat{\mathcal{A}}$ for $\lambda_n = \frac{1}{\rho_n} \in (0, 1)$ and further $\|x_n - \hat{z}_0\| = 1$ for all $n \geq n_0$. In Corollary 3.6 we have shown the α-uniform convexity of $\hat{\mathcal{E}}(t,\cdot)$ and we deduce

$$\lambda_n \hat{\mathcal{E}}(t, \hat{z}_0) + (1 - \lambda_n)\hat{\mathcal{E}}(t, z_n) \geq \hat{\mathcal{E}}(t, x_n) + \frac{\alpha}{2}\lambda_n(1 - \lambda_n)\|z_n - \hat{z}_0\|^2$$

$$= \hat{\mathcal{E}}(t, x_n) + \frac{\alpha}{2}\big(\|z_n - \hat{z}_0\| - 1\big).$$

Due to the continuity of $\hat{\mathcal{E}}(t,\cdot)$ (see Assumption (3.2)) we have boundedness of $\hat{\mathcal{E}}(t,\cdot)$ on the compact set $\left\{z \in \hat{\mathcal{A}} : \|z - \hat{z}_0\| = 1\right\}$ and there exists a constant $c_{\min} > 0$ such that $\hat{\mathcal{E}}(t, z) \geq -c_{\min}$ for all $\|z - \hat{z}_0\| = 1$.
Due to this and due to $\lambda_n \in (0, 1)$ we deduce from above that

$$\hat{\mathcal{E}}(t, \hat{z}_0) + \hat{\mathcal{E}}(t, z_n) \geq -c_{\min} - \frac{\alpha}{2} + \frac{\alpha}{2}\|z_n - \hat{z}_0\|.$$

The coercivity follows directly from the last estimate.   ∎

Definition 3.9 (Minimizing problem) *For given $\tau, t \in [\hat{T}, \hat{T} + \hat{\Delta}]$ and $x, w \in \hat{\mathcal{A}}$ find*

$$\hat{z} \in \operatorname{argmin}\left\{\hat{\mathcal{E}}(\tau, y) + \hat{\Psi}(t, x, y - w) : y \in \hat{\mathcal{A}}\right\}. \tag{MP}$$

Here "argmin" denotes the set of all minimizers.

Proposition 3.10 (Existence & uniqueness for (MP)**)** *Let us assume* (3.1)–(3.6) *then there exists for given* $\tau, t \in [\hat{T}, \hat{T} + \hat{\Delta}]$ *and* $x, w \in \hat{\mathcal{A}}$ *a unique solution* $\hat{z} \in \hat{\mathcal{A}}$ *of the Minimizing Problem* (MP).

Proof: For given $\tau, t \in [\hat{T}, \hat{T} + \hat{\Delta}]$, $x, w \in \hat{\mathcal{A}}$ the functional $\hat{J}(y) := \hat{\mathcal{E}}(\tau, y) + \hat{\Psi}(t, x, y - w)$ is coercive, since $\hat{\Psi} \geq 0$ and $\hat{\mathcal{E}}(\tau, \cdot)$ is also coercive (see Lemma 3.7). The regularity Assumptions (3.2) and (3.4) on $\hat{\mathcal{E}}$ and $\hat{\Psi}$ imply the continuity of $\hat{J}$. The existence of a minimizer $\hat{z} \in \hat{\mathcal{A}}$ follows from Weierstraß' extrem principle. The convexity of the functional $\hat{\Psi}$ (see Lemma 3.3) together with the α-uniform convexity of $\hat{\mathcal{E}}$ imply the strict convexity of $\hat{J}$. Together with the convexity of the set $\mathcal{A}$ this proves uniqueness.   ∎

Definition 3.11 (Incremental Problem and approximative solutions) *For a given initial value* $z_0 = \hat{z}_0 \in \hat{\mathcal{A}}$, *the time interval* $[\hat{T}, \hat{T} + \hat{\Delta}]$ *and the equi-distant partition* $\Pi : \hat{T} = t_0 < t_1 < \cdots < t_{N_\Pi} = \hat{T} + \hat{\Delta}$ *with fineness* $f_\Pi = \hat{\Delta}/N_\Pi$ *the Incremental Problem consists in finding incrementally* $z_1, z_2, z_3, \ldots, z_{N_\Pi}$ *such that for* $k = 1, \ldots, N_\Pi$

$$z_k \in \arg\min \left\{ \hat{\mathcal{E}}(t_k, y) + \hat{\Psi}(t_{k-1}, z_{k-1}, y - z_{k-1}) \; : \; y \in \hat{\mathcal{A}} \right\}. \tag{IP}$$

For a sequence of equi-distant partitions $(\Pi^{(n)})_{n \in \mathbb{N}}$ *of* $[\hat{T}, \hat{T} + \hat{\Delta}]$, *i.e.*

$$\Pi^{(n)} : \hat{T} = t_0^{(n)} < t_1^{(n)} < \cdots < t_{N_{\Pi^{(n)}}}^{(n)} = \hat{T} + \hat{\Delta}$$

we define the sequence of approximative solutions $(z^{(n)})_{n \in \mathbb{N}} \subset \mathrm{W}^{1,\infty}([\hat{T}, \hat{T} + \hat{\Delta}], \hat{\mathcal{A}})$ *as the piecewise linear interpolation between the values* $(z_k^{(n)})_{k=1,\ldots,N_{\Pi^{(n)}}}$ *of the solution of the corresponding Incremental Problem* (IP), *i.e.*

$$z^{(n)}(t) := z_{k-1}^{(n)} + \frac{t - t_{k-1}^{(n)}}{t_k^{(n)} - t_{k-1}^{(n)}} (z_k^{(n)} - z_{k-1}^{(n)}) \quad \text{for } t \in [t_{k-1}^{(n)}, t_k^{(n)}] \text{ and } k = 1, \ldots, N_{\Pi^{(n)}}.$$

Lemma 3.12 (Existence and uniqueness) *Let us assume* (3.1)–(3.6) *and an initial value* $\hat{z}_0 \in \hat{\mathcal{A}}$ *and a time interval* $[\hat{T}, \hat{T} + \hat{\Delta}]$ *to be given. Then there exists for each partition* Π *of* $[\hat{T}, \hat{T} + \hat{\Delta}]$ *a unique solution* $(z_k)_{k=1,\ldots,N_\Pi}$ *in* $\hat{\mathcal{A}}$ *of the incremental Problem 3.11.*
Further there exists for each sequence of partitions $(\Pi^{(n)})_{n \in \mathbb{N}}$ *of* $[\hat{T}, \hat{T} + \hat{\Delta}]$ *a unique sequence of approximative solutions* $(z^{(n)})_{n \in \mathbb{N}} \subset \mathrm{W}^{1,\infty}([\hat{T}, \hat{T} + \hat{\Delta}], \hat{\mathcal{A}})$.

Proof: Since we solve incrementally Minimizing Problems (MP) the results follow directly from Proposition 3.10.   ∎

Remark 3.13 *Note that the Definition 3.11 of the approximative solutions holds for an arbitrary sequence of equi-distant partitions $(\Pi^{(n)})_{n \in \mathbb{N}}$. Of course we are interested in a sequence of partitions such that the fineness tends to 0, i.e.*

$$f_{\Pi^{(n)}} := \frac{\hat{\Delta}}{N_{\Pi^{(n)}}} \to 0 \quad \text{for } n \to \infty. \tag{3.17}$$

3.5   Convergence

In this section we prove the existence of a convergent subsequence of the approximative solutions $(z^{(n)})_{n \in \mathbb{N}} \subset W^{1,\infty}\left([\hat{T}, \hat{T} + \hat{\Delta}], \hat{\mathcal{A}}\right)$ which we introduced in Definition 3.11. For the convergence result we will need the additional Assumptions (3.7)–(3.11), the final result is Theorem 3.20. The whole proof is based on the Arzela-Ascoli theorem.

3.5.1   Arzela-Ascoli

Theorem 3.14 (Arzela-Ascoli) *Let $\mathcal{S} \subset \mathbb{R}^m$ be a compact set and assume that the sequence $(f_j)_{j \in \mathbb{N}} \subset C^0(\mathcal{S}, \mathbb{R}^m)$ is bounded and equicontinuous, i.e.*

$$\sup\left\{\|f_j\|_{C^0(\mathcal{S}, \mathbb{R}^m)} \; : \; j \in \mathbb{N}\right\} \leq C \quad and$$
$$\sup\left\{\|f_j(y) - f_j(x)\| \; : \; j \in \mathbb{N}\right\} \to 0 \quad for \quad \|y - x\| \to 0,$$

then there exists a convergent subsequence $(f_{j_k})_{k \in \mathbb{N}} \subset (f_j)_{j \in \mathbb{N}}$ and a limit function $f \in C^0(\mathcal{S}, \mathbb{R}^m)$ such that

$$\|f_{j_k} - f\|_{C^0(\mathcal{S}, \mathbb{R}^m)} \to 0 \quad for \; k \to \infty.$$

Since we want to apply the Arzela-Ascoli theorem to our sequence of approximative solutions $(z^{(n)})_{n \in \mathbb{N}} \subset W^{1,\infty}([\hat{T}, \hat{T} + \hat{\Delta}], \hat{\mathcal{A}})$ defined in 3.11 we can exploit that all functions satisfy the same initial condition, i.e. $z^{(n)}(\hat{T}) = \hat{z}_0$ for all $n \in \mathbb{N}$. This allows us to simplify the assumptions of the *Arzela-Ascoli*-theorem as follows.

Theorem 3.15 (Unif. Lip.- continuity implies convergent subsequence) *Assume that the sequence of functions $(z^{(n)})_{n \in \mathbb{N}} \subset W^{1,\infty}([\hat{T}, \hat{T} + \hat{\Delta}], \hat{\mathcal{A}})$ satisfies*

$$z^{(n)}(\hat{T}) = \hat{z}_0 \quad for \; all \; n \in \mathbb{N}$$

and that there exists a uniform Lipschitz constant $c_{\text{lip}} > 0$, i.e.

$$\|z^{(n)}(t_2) - z^{(n)}(t_1)\| \leq c_{\text{lip}}|t_2 - t_1| \quad for \; all \; t_1, t_2 \in [\hat{T}, \hat{T} + \hat{\Delta}] \; and \; n \in \mathbb{N}, \tag{3.18}$$

then there exists a convergent subsequence $(z^{(n_j)})_{j \in \mathbb{N}}$ and a limit function $z \in W^{1,\infty}([\hat{T}, \hat{T} + \hat{\Delta}], \hat{\mathcal{A}})$ such that

$$\|z^{(n_j)} - z\|_{C^0([\hat{T}, \hat{T} + \hat{\Delta}], \hat{\mathcal{A}})} \to 0 \quad for \; j \to \infty.$$

Proof: Since $[\hat{T}, \hat{T} + \hat{\Delta}] \subset \mathbb{R}$ is compact, it is enough to prove the boundedness and equicontinuity of the sequence $(z^{(n)})_{n \in \mathbb{N}} \subset W^{1,\infty}([\hat{T}, \hat{T} + \hat{\Delta}], \hat{\mathcal{A}})$. Due to the initial condition and the uniform Lipschitz continuity (3.18) we have

$$\|z^{(n)}(t)\| \leq \|\hat{z}_0\| + c_{\mathrm{lip}}\hat{\Delta} =: \mathrm{C}.$$

Note that $\mathrm{C} > 0$ is independent of $n \in \mathbb{N}$ and $t \in [\hat{T}, \hat{T} + \hat{\Delta}]$. Hence the sequence is bounded. The equicontinuity follows directly form the uniform Lipschitz continuity.
∎

Since the definition of the approximative solutions 3.11 is directly based on the solutions $(z_k^{(n)})_{k=1,\ldots,N_{\Pi^{(n)}}}$ of the corresponding Incremental Problems (IP) it is sufficient to prove the existence of $c_{\mathrm{lip}} > 0$ independent of $n \in \mathbb{N}$ such that

$$\|z_k^{(n)} - z_{k-1}^{(n)}\| \leq c_{\mathrm{lip}}|t_k^{(n)} - t_{k-1}^{(n)}| \tag{3.19}$$

holds for $k = 1, \ldots, N_{\Pi^{(n)}}$ and all $n \in \mathbb{N}$.

Hence, our remaining task is to prove the existence of a Lipschitz constant $c_{\mathrm{lip}} > 0$ that is independent of any equi-distant partition such that for any solution of the Incremental Problem (IP) the estimate (3.19) holds. Since solutions of (IP) are constructed by incrementally solving Minimizing Problems (MP) we derive in the following section some useful estimates for solutions of (MP).

3.5.2   Recursive estimates for solutions

Lemma 3.16 (First estimate for (MP)) *Let the times $\tau, t \in [\hat{T}, \hat{T} + \hat{\Delta}]$ and the values $x, w \in \hat{\mathcal{A}}$ be given. We assume (3.1)–(3.6) then the solution $z \in \hat{\mathcal{A}}$ of the Minimizing Problem*

$$z \in \operatorname{argmin}\left\{ \hat{\mathcal{E}}(\tau, y) + \hat{\Psi}(t, x, y - w) \ : \ y \in \hat{\mathcal{A}} \right\} \tag{MP}$$

satisfies

$$\int_0^1 \int_0^r \langle y - z, \hat{\mathrm{H}}(\tau, y(s))(y-z)\rangle \mathrm{d}s \mathrm{d}r \leq \hat{\mathcal{E}}(\tau, y) - \hat{\mathcal{E}}(\tau, z) + \hat{\Psi}(t, x, y-w) - \hat{\Psi}(t, x, z-w)$$

for all $y \in \hat{\mathcal{A}}$, where the Hessian matrix $\hat{\mathrm{H}}(t, z) \in \mathbb{R}^{3N \times 3N}$ was introduced in (3.3) and $y(s) := z + s(y - z) \in \hat{\mathcal{A}}$.

Proof: We split the proof into two parts. First we consider the difference $\hat{\mathcal{E}}(\tau, y) - \hat{\mathcal{E}}(\tau, z)$ and apply twice the fundamental theorem of calculus.

$$
\begin{aligned}
\hat{\mathcal{E}}(\tau, y) - \hat{\mathcal{E}}(\tau, z) &= \int_0^1 \left(\frac{\mathrm{d}}{\mathrm{d}r} \hat{\mathcal{E}}(\tau, y(r)) \right)(r)\mathrm{d}r \\
&= \int_0^1 \mathrm{D}\hat{\mathcal{E}}(\tau, y(r))(y - z)\mathrm{d}r \\
&= \int_0^1 \mathrm{D}\hat{\mathcal{E}}(\tau, y(r)) - \mathrm{D}\hat{\mathcal{E}}(\tau, y(0))(y - z)\mathrm{d}r + \int_0^1 \mathrm{D}\hat{\mathcal{E}}(\tau, z)(y - z)\mathrm{d}r \\
&= \int_0^1 \int_0^r \langle y - z, \hat{\mathrm{H}}(\tau, y(s))(y - z)\rangle \mathrm{d}s\mathrm{d}r + \mathrm{D}\hat{\mathcal{E}}(\tau, z)(y - z).
\end{aligned}
$$

In the second part we reformulate the abstract minimizing problem using the characteristic function $\mathcal{X}_{\hat{A}}(z) = \begin{cases} 0 & z \in \hat{A} \\ +\infty & \text{else} \end{cases}$ as follows

$$
z = \operatorname{argmin} \left\{ \hat{\mathcal{E}}(\tau, y) + \hat{\Psi}(t, x, y - w) + \mathcal{X}_{\hat{A}}(y) \ : \ y \in \mathbb{R}^{3N} \right\}.
$$

Denoting by ∂_y the subdifferential with respect to y this is equivalent to

$$
-\mathrm{D}\hat{\mathcal{E}}(\tau, z) \in \partial_y \left\{ \hat{\Psi}(t, x, \cdot - w) + \mathcal{X}_{\hat{A}}(\cdot) \right\}(z).
$$

Here we used the convexity of $\mathcal{E}(t, \cdot)$ (see Corollary 3.6) and the *Moreau-Rockafellar-Theorem A.3*. By definition of the subdifferential this is equivalent to

$$
\begin{aligned}
-\mathrm{D}\hat{\mathcal{E}}(\tau, z)(y - z) &\le \hat{\Psi}(t, x, y - w) - \hat{\Psi}(t, x, z - w) + \mathcal{X}_{\hat{A}}(y) - \mathcal{X}_{\hat{A}}(z) \\
&= \hat{\Psi}(t, x, y - w) - \hat{\Psi}(t, x, z - w).
\end{aligned}
$$

The first and the second part prove together the lemma. ∎

Lemma 3.17 (Second estimate for (MP)) *Let the times* $\tau_1, \tau_2, t_1, t_2 \in [\hat{T}, \hat{T} + \hat{\Delta}]$ *and the value* $z_0 \in \hat{A}$ *be given. We assume* (3.1)–(3.6) *and define first* $z_1 \in \hat{A}$ *and then* $z_2 \in \hat{A}$ *via*

$$
z_1 \in \operatorname{argmin} \left\{ \hat{\mathcal{E}}(\tau_1, y) + \hat{\Psi}(t_1, z_0, y - z_0) \ : \ y \in \hat{A} \right\} \quad \text{and} \tag{3.20}
$$

$$
z_2 \in \operatorname{argmin} \left\{ \hat{\mathcal{E}}(\tau_2, y) + \hat{\Psi}(t_2, z_1, y - z_1) \ : \ y \in \hat{A} \right\}. \tag{3.21}
$$

Then the estimate

$$
\int_0^1 \int_0^r \langle z_2 - z_1, \left(\hat{\mathrm{H}}(\tau_1, y(s)) + \hat{\mathrm{H}}(\tau_2, y(1-s)) \right)(z_2 - z_1)\rangle \mathrm{d}s\,\mathrm{d}r \tag{3.22}
$$

$$
\le \int_0^1 \int_0^r \partial_t \mathrm{D}\hat{\mathcal{E}}(\tau(r), y(s))(z_2 - z_1)\,\mathrm{d}s\,\mathrm{d}r(\tau_1 - \tau_2) + \hat{\Psi}(t_1, z_0, z_2 - z_1) - \hat{\Psi}(t_2, z_1, z_2 - z_1).
$$

holds where $y(s) := z_1 + s(z_2 - z_1)$ *and* $\tau(r) := \tau_1 + r(\tau_2 - \tau_1)$.

Proof: We apply Lemma 3.16 twice, once for (3.20) with test value $y = z_2 \in \hat{\mathcal{A}}$ and once for (3.21) with the choice $y = z_1 \in \hat{\mathcal{A}}$. Note that we use different times τ_j and t_j for $j = 1, 2$. Adding the two resulting estimates leads us to

$$\int_0^1 \int_0^r \langle z_2 - z_1, \left(\hat{\mathrm{H}}(\tau_1, y(s)) + \hat{\mathrm{H}}(\tau_2, y(1-s)) \right) (z_2 - z_1) \rangle \, \mathrm{d}s \, \mathrm{d}r$$

$$\leq \hat{\mathcal{E}}(\tau_1, z_2) - \hat{\mathcal{E}}(\tau_1, z_1) + \hat{\mathcal{E}}(\tau_2, z_1) - \hat{\mathcal{E}}(\tau_2, z_2)$$

$$+ \hat{\Psi}(t_2, z_1, z_1 - z_1) - \hat{\Psi}(t_2, z_1, z_2 - z_1) + \hat{\Psi}(t_1, z_0, z_2 - z_0) - \hat{\Psi}(t_1, z_0, z_1 - z_0).$$

The left hand side is already as desired while for the right hand side we apply the fundamental theorem of calculus twice to see

$$\sum_{j,k=1}^2 (-1)^{j+k+1} \hat{\mathcal{E}}(\tau_j, z_k) = \int_0^1 \int_0^1 \partial_t \mathrm{D} \hat{\mathcal{E}}(\tau(r), y(s))(z_2 - z_1) \, \mathrm{d}s \, \mathrm{d}r (\tau_1 - \tau_2).$$

Further we observe that due to the homogeneity of degree one of $\hat{\Psi}(t, x, \cdot)$ (see (3.5)) we have $\hat{\Psi}(t_2, z_1, z_1 - z_1) = \hat{\Psi}(t_2, z_1, 0) = 0$. It remains us to estimate

$$\hat{\Psi}(t_1, z_0, z_2 - z_0) - \hat{\Psi}(t_1, z_0, z_1 - z_0) \leq \hat{\Psi}(t_1, z_0, z_2 - z_1).$$

But this is exactly the triangle inequality that we assumed in (3.6) and the proof is complete.   ∎

Up to now we assumed (3.1)–(3.6) only and didn't use the continuity of $\hat{\Psi} : [\hat{T}, \hat{T} + \hat{\Delta}] \times \hat{\mathcal{A}} \times \mathbb{R}^{3N} \to [0, \infty)$ with respect to its second variable $z \in \hat{\mathcal{A}}$. But the following lemma really relies on the (Lipschitz-) continuity of $\hat{\Psi}$ with respect to $z \in \hat{\mathcal{A}}$ and hence avoids all the difficulties that will arise from contact and non-contact modeling.

Lemma 3.18 (Third estimate for (MP)**)** *Let the times* $\tau_1, \tau_2, t_1, t_2 \in [\hat{T}, \hat{T} + \hat{\Delta}]$ *and the value* $\hat{z}_0 \in \hat{\mathcal{A}}$ *be given. We assume* (3.1)–(3.9) *and define first* $z_1 \in \hat{\mathcal{A}}$ *and then* $z_2 \in \hat{\mathcal{A}}$ *by*

$$z_1 \in \mathrm{argmin} \left\{ \hat{\mathcal{E}}(\tau_1, y) + \hat{\Psi}(t_1, \hat{z}_0, y - \hat{z}_0) : y \in \hat{\mathcal{A}} \right\} \quad and \quad (3.23)$$

$$z_2 \in \mathrm{argmin} \left\{ \hat{\mathcal{E}}(\tau_2, y) + \hat{\Psi}(t_2, z_1, y - z_1) : y \in \hat{\mathcal{A}} \right\}. \quad (3.24)$$

Then the following recursive estimate holds

$$\hat{\alpha}_* \| z_2 - z_1 \| \leq (\hat{c}_1 + \hat{c}_2) \max\{ |\tau_2 - \tau_1|, |t_2 - t_1| \} + \hat{q} \| z_1 - \hat{z}_0 \|. \quad (3.25)$$

Proof: For $z_2 = z_1$ there is nothing to prove. Hence we assume $\| z_2 - z_1 \| > 0$ in the following. We make use of the estimate (3.22) of the previous Lemma 3.17. With the help of the convexity Assumption (3.3) on $\hat{\mathcal{E}}(\tau, \cdot)$ we estimate the double integral on the left hand side of (3.22) from below by $\hat{\alpha}_*\|z_2 - z_1\|^2$ while the double integral on the right side is estimated from above with the help of (3.7) by $\hat{c}_1\|z_2 - z_1\|\,|\tau_2 - \tau_1|$. This leads us to

$$\hat{\alpha}_*\|z_2 - z_1\|^2 \le \hat{c}_1\|z_2 - z_1\|\,|\tau_2 - \tau_1| + \hat{\Psi}(t_1, \hat{z}_0, z_2 - z_1) - \hat{\Psi}(t_2, z_1, z_2 - z_1).$$

Next we make use of $\|z_2 - z_1\| > 0$ and we define $v := \dfrac{z_2 - z_1}{\|z_2 - z_1\|} \in \mathbb{S}^{3N-1}$. With the help of this notation and since $\hat{\Psi}$ is homogeneous of degree one with respect to its third variable we derive

$$\hat{\Psi}(t_1, \hat{z}_0, z_2 - z_1) - \hat{\Psi}(t_2, z_1, z_2 - z_1) = \left(\hat{\Psi}(t_1, \hat{z}_0, v) - \hat{\Psi}(t_2, z_1, v) \right) \|z_2 - z_1\|.$$

Adding the Assumptions (3.8)–(3.9) we get

$$\left(\hat{\Psi}(t_1, \hat{z}_0, v) - \hat{\Psi}(t_2, z_1, v) \right) \le \hat{c}_2|t_2 - t_1| + \hat{q}\|z_1 - \hat{z}_0\|.$$

We summarize now the last three estimates to

$$\hat{\alpha}_*\|z_2 - z_1\|^2 \le (\hat{c}_1 + \hat{c}_2)\max\{|t_2 - t_1|, |\tau_2 - \tau_1|\}\|z_2 - z_1\| + \hat{q}\|z_1 - \hat{z}_0\|\|z_2 - z_1\|.$$

After dividing both sides with $\|z_2 - z_1\|$ we complete our proof. ∎

3.5.3 Lipschitz continuity & convergence

Theorem 3.19 (Lipschitz continuity for (IP)**)** *Let* (3.1)–(3.11) *hold. Then the solution* $(z_k)_{k=1,\dots,N_\Pi}$ *of the corresponding Incremental Problem* (IP) *exists and satisfies estimate* (3.18), *i.e.*

$$\|z_k - z_{k-1}\| \le c_{\text{lip}}|t_k - t_{k-1}| \quad for\ k = 1, \dots, N_\Pi \tag{3.26}$$

with the choice $c_{\text{lip}} := \dfrac{\hat{c}_1 + \hat{c}_2}{\hat{\alpha}_* - \hat{q}}$ *and with the constants* $\hat{\alpha}_*, \hat{c}_1, \hat{c}_2$ *and* $\hat{q}$ *defined in* (3.3) *and* (3.7)–(3.9) *and being independent of the equi-distant partition* Π *of* $[\hat{T}, \hat{T} + \hat{\Delta}]$.

Proof: The existence follows from Lemma 3.12. For the Lipschitz continuity we note that due to the stability Assumption (3.11) the initial value $\hat{z}_0$ satisfies

$$\hat{z}_0 \in \operatorname{argmin}\left\{ \hat{\mathcal{E}}(\hat{T}, y) + \hat{\Psi}(\hat{T}, \hat{z}_0, y - \hat{z}_0) \;:\; y \in \hat{\mathcal{A}} \right\}$$

while z_1 is defined by

$$z_1 \in \operatorname{argmin}\left\{ \hat{\mathcal{E}}(t_1, y) + \hat{\Psi}(\hat{T}, \hat{z}_0, y - \hat{z}_0) \;:\; y \in \hat{\mathcal{A}} \right\}.$$

Now we are exactly in the setting of the previous Lemma 3.18. There, we find $t_1 = t_2 = \hat{T}, \tau_1 = \hat{T}, \tau_2 = t_1$ and $\hat{z}_0 \equiv z_1$ in the estimate (3.25) and we deduce for our current notation here that

$$\|z_1 - \hat{z}_0\| \leq \frac{\hat{c}_1 + \hat{c}_2}{\hat{\alpha}_*}|t_1 - \hat{T}| \quad \text{holds.} \tag{3.27}$$

For $k = 2, \ldots, N_\Pi$ the values z_{k-1} and z_k are defined via

$$z_{k-1} \in \operatorname{argmin}\left\{\hat{\mathcal{E}}(t_{k-1}, y) + \hat{\Psi}(t_{k-2}, z_{k-2}, y - z_{k-2}) : y \in \hat{\mathcal{A}}\right\} \quad \text{and}$$

$$z_k \in \operatorname{argmin}\left\{\hat{\mathcal{E}}(t_k, y) + \hat{\Psi}(t_{k-1}, z_{k-1}, y - z_{k-1}) : y \in \hat{\mathcal{A}}\right\}$$

and again we use Lemma 3.18 to estimate

$$\hat{\alpha}_*\|z_k - z_{k-1}\| \leq (\hat{c}_2 + \hat{c}_1)f_\Pi + \hat{q}\|z_{k-1} - z_{k-2}\|$$

for $k = 2, \ldots, N_\Pi$ and fineness $f_\Pi = t_k - t_{k-1} = \frac{1}{N_\Pi}$. We now exploit the last inequality recursively to get, for arbitrary $k = 2, \ldots, N_\Pi$,

$$\|z_k - z_{k-1}\| \leq \frac{\hat{c}_1 + \hat{c}_2}{\hat{\alpha}_*}f_\Pi \sum_{j=0}^{k-1}\left(\frac{\hat{q}}{\hat{\alpha}_*}\right)^j + \left(\frac{\hat{q}}{\hat{\alpha}_*}\right)^k\|z_1 - \hat{z}_0\|.$$

We let $r = \left(\dfrac{\hat{q}}{\hat{\alpha}_*}\right)$. Using (3.27) to estimate $\|z_1 - \hat{z}_0\|$, this leads us directly to

$$\|z_k - z_{k-1}\| \leq \frac{\hat{c}_1 + \hat{c}_2}{\hat{\alpha}_*}f_\Pi \sum_{j=0}^{k-1} r^j. \tag{3.28}$$

Since (3.10) implied $r < 1$ it is easy to see that

$$\sum_{j=0}^{k-1} r^j = \frac{1 - r^k}{1 - r} \leq \frac{1}{1 - r} = \frac{\hat{\alpha}_*}{\hat{\alpha}_* - \hat{q}}$$

holds. Together with (3.28) and the definition $c_{\text{lip}} := \frac{\hat{c}_1 + \hat{c}_2}{\hat{\alpha}_* - \hat{q}}$ this proves $\|z_k - z_{k-1}\| \leq c_{\text{lip}}f_\Pi$ (3.26) also for $k = 2, \ldots, N_\Pi$. Thus our proof is complete. ∎

We now summarize all results to prove

Theorem 3.20 (Convergence) *Let the Assumptions (3.1)–(3.11) hold and assume a sequence of equi-distant partitions $\Pi^{(n)} : \hat{T} = \tau_0^{(n)} < \cdots < \tau_{N_{\Pi^{(n)}}}^{(n)} = \hat{T} + \hat{\Delta}$ to be given then the corresponding sequence of approximative solutions $(z^{(n)})_{n \in \mathbb{N}} \subset \mathrm{W}^{1,\infty}([\hat{T}, \hat{T}+\hat{\Delta}], \hat{\mathcal{A}})$ has a convergent subsequence $(z^{(n_j)})_{j \in \mathbb{N}} \subset \mathrm{W}^{1,\infty}([\hat{T}, \hat{T}+\hat{\Delta}], \hat{\mathcal{A}})$, i.e.*

$$\sup\left\{\|z^{(n_j)}(t) - z(t)\| : t \in [\hat{T}, \hat{T} + \hat{\Delta}]\right\} \to 0 \quad \text{for } j \to \infty. \tag{3.29}$$

The limit function z satisfies for almost all $t \in [\hat{T}, \hat{T} + \hat{\Delta}]$ the Lipschitz estimate

$$\|\dot{z}(t)\| \leq \frac{\hat{c}_1 + \hat{c}_2}{\hat{\alpha}_* - \hat{q}} \tag{3.30}$$

with the constants $\hat{\alpha}_*, \hat{c}_1, \hat{c}_2$ and $\hat{q}$ defined in (3.3) and (3.7)–(3.9).

Proof: Theorem 3.15 shows that the uniform Lipschitz continuity of the sequence $(z^{(n)})_{n \in \mathbb{N}}$
$\subset \mathrm{W}^{1,\infty}([\hat{T}, \hat{T} + \hat{\Delta}], \hat{\mathcal{A}})$ of approximative solutions implies the existence of a convergent subsequence. The Definition 3.11 of the approximative solutions is based on the corresponding sequence of incremental solutions $\left((z^{(n)})_{k=0,\ldots,N_\Pi^{(n)}} \right)_{n \in \mathbb{N}}$ and the uniform Lipschitz continuity follows from the estimate

$$\|z_k^{(n)} - z_{k-1}^{(n)}\| \leq c_{\mathrm{lip}} |t_k^{(n)} - t_{k-1}^{(n)}| \quad \text{for all } k = 1, \ldots, N_\Pi^{(n)} \text{ and } n \in \mathbb{N}$$

if the constant $c_{\mathrm{lip}} > 0$ is independent of k and n. But our previous Theorem 3.19 shows exactly this estimate with the choice $c_{\mathrm{lip}} = \frac{\hat{c}_1 + \hat{c}_2}{\hat{\alpha}_* - \hat{q}}$. Note that the involved constants are independent of $n \in \mathbb{N}$ and thus c_{lip} is independent, too. ∎

3.6 Existence

In this section we want to prove that the limit function $z \in \mathrm{W}^{1,\infty}([\hat{T}, \hat{T} + \hat{\Delta}], \hat{\mathcal{A}})$ that we found in the previous Section 3.5 is a solution of the Energetic Problem 3.1. We follow ideas of [MiR07] and only have to be careful when we only assume the continuity of dissipation $\hat{\Psi}(\cdot, z\,\cdot)$. We want to reuse and recall the results of this *Existence* Section later on again. That is why we now make clear what the starting point of this section is and which are the minimal assumptions.

Let a convex set $\hat{\mathcal{A}} \subset \mathbb{R}^{3N}$, an energy functional $\hat{\mathcal{E}} : [\hat{T}, \hat{T} + \hat{\Delta}] \times \hat{\mathcal{A}} \to \mathbb{R}$, a dissipation functional $\hat{\Psi} : [\hat{T}, \hat{T} + \hat{\Delta}] \times \hat{\mathcal{A}} \times \mathbb{R}^{3N} \to [0, \infty)$ and an initial value $\hat{z}_0 \in \hat{\mathcal{A}}$ be given. We choose a sequence of equi-distant partitions $(\Pi^{(n)})_{n \in \mathbb{N}}$ of $[\hat{T}, \hat{T} + \hat{\Delta}]$ whose fineness, as defined in (3.17), vanishes, i.e.

$$f(\Pi^{(n)}) \to 0 \quad \text{for } n \to \infty.$$

We assume that the corresponding sequence of Incremental Problems 3.11 has a sequence of (unique) discrete solutions

$$(z_{k=1,\ldots,N_{\Pi^{(n)}}}^{(n)})_{n \in \mathbb{N}}. \tag{3.31}$$

Further we assume the existence of a constant $c_{\mathrm{lip}} > 0$ which is independent of $n, k \in \mathbb{N}$ such that

$$\|z_k^{(n)} - z_{k-1}^{(n)}\| \leq c_{\mathrm{lip}} |t_k - t_{k-1}| \tag{3.32}$$

holds for all $k = 1, \ldots, N_{\Pi^{(n)}}$ and all $n \in \mathbb{N}$.

Finally we make the assumption that there exists some **limit function**

$$z \in W^{1,\infty}([\hat{T}, \hat{T} + \hat{\Delta}], \hat{\mathcal{A}})$$

and a convergent subsequence (still denoted by $(z^{(n)})_{n \in \mathbb{N}} \subset W^{1,\infty}([\hat{T}, \hat{T} + \hat{\Delta}], \hat{\mathcal{A}})$) of the approximative solutions, as they were defined in 3.11 such that

$$\|z^{(n)} - z\|_{C^0([\hat{T}, \hat{T} + \hat{\Delta}], \hat{\mathcal{A}})} \to 0 \quad \text{for } n \to \infty. \tag{3.33}$$

The major simplification of this chapter is that we assumed Lipschitz continuity of the dissipation $\hat{\Psi}$ with respect to z in (3.9). We needed this assumption to construct convergent subsequences in the previous section but we want to use this strong assumption as little as possible and avoid it in this section. Anyhow, proving existence by passing to the limit we will need some kind of continuity of $\hat{\Psi}$ with respect to z. The following assumption is a lot weaker than (3.9). We claim the existence of a closed subset $\mathcal{D} \subset [\hat{T}, \hat{T} + \hat{\Delta}] \times \hat{\mathcal{A}}$ such that

$$\begin{aligned} &\hat{\Psi} \in C\left(\mathcal{D} \times \mathbb{R}^{3N}, \mathbb{R}\right) \quad \text{and} \\ &(t_k, z_k)_{k=1,\ldots,N_\Pi} \subset \mathcal{D} \quad \text{for solutions of (IP).} \end{aligned} \tag{3.34}$$

This is the only assumption which does not follow as a result from the previous section.

Our remaining task is to show under these assumptions that a limit function $z \in W^{1,\infty}\left([\hat{T}, \hat{T} + \hat{\Delta}], \hat{\mathcal{A}}\right)$ provides us with a solution of Problem 3.1. Hence the function z has to satisfy the initial condition $z(\hat{T}) = \hat{z}_0$, the stability condition (S) and the energy equality (E).

3.6.1   Initial condition

Since all approximative solutions satisfy $z^{(n)}(\hat{T}) = \hat{z}_0$ the Convergence Assumption (3.33) gives

$$z(\hat{T}) = \hat{z}_0. \tag{3.35}$$

3.6.2   Stability

We constructed the approximative solutions with the help of discrete solutions $(z_k)_{k=1,\ldots,N_\Pi}$ defined by the Incremental Problem 3.11. We now show that these discrete solutions satisfy a discrete version of the stability property (S).

Lemma 3.21 (Discrete stability for (IP)) *Let us assume that $(z_k)_{k=1,\ldots,N_\Pi}$ is a solution of the Incremental Problem 3.11 and that the triangle inequality (3.6) for $\hat{\Psi}$ holds, then a discrete stability condition (S_{dis}) holds for each $k = 1, \ldots, N_\Pi$, i.e.*

$$\hat{\mathcal{E}}(t_k, z_k) \leq \hat{\mathcal{E}}(t_k, y) + \hat{\Psi}(t_{k-1}, z_{k-1}, y - z_k) \quad \text{for all } y \in \hat{\mathcal{A}}. \tag{S_{dis}}$$

Proof: Since $(z_k)_{k=1,\ldots,N_\Pi}$ is a solution of (IP) we get for each $k = 1, \ldots, N_\Pi$

$$\hat{\mathcal{E}}(t_k, z_k) + \hat{\Psi}(t_{k-1}, z_{k-1}, z_k - z_{k-1}) \leq \hat{\mathcal{E}}(t_k, y) + \hat{\Psi}(t_{k-1}, z_{k-1}, y - z_{k-1}) \text{ for all } y \in \hat{\mathcal{A}}.$$

The triangle inequality for $\hat{\Psi}$ claimed in (3.6) supplies us with

$$\hat{\Psi}(t_{k-1}, z_{k-1}, y - z_{k-1}) \leq \hat{\Psi}(t_{k-1}, z_{k-1}, y - z_k) + \hat{\Psi}(t_{k-1}, z_{k-1}, z_k - z_{k-1}).$$

Both inequalities together establish the result. ∎

Lemma 3.22 (Stability of the limit function z) *Let the Assumptions* (3.31)–(3.34), *the Regularity Assumption* (3.2) *on $\hat{\mathcal{E}}$ and the triangle inequality* (3.6) *for $\hat{\Psi}$ hold, then the limit function z is stable for all $t \in [\hat{T}, \hat{T} + \hat{\Delta}]$, i.e.*

$$\hat{\mathcal{E}}(t, z(t)) \leq \hat{\mathcal{E}}(t, y) + \hat{\Psi}(t, z(t), y - z(t)) \quad \text{for all } y \in \hat{\mathcal{A}}. \tag{S}$$

Proof: Each discrete solution $(z_k^{(n)})_{k=1,\ldots,N_{\Pi^{(n)}}} \subset \hat{\mathcal{A}}$ satisfies due to Assumption (3.6) and the previous Lemma 3.21 the discrete stability (S_{dis}), i.e.

$$\hat{\mathcal{E}}(\tau_k^{(n)}, z_k^{(n)}) \leq \hat{\mathcal{E}}(\tau_k^{(n)}, y) + \hat{\Psi}(\tau_{k-1}^{(n)}, z_{k-1}^{(n)}, y - z_{k-1}^{(n)}) \quad \text{for all } y \in \hat{\mathcal{A}} \tag{3.36}$$

and for all $k = 1, \ldots, N_{\Pi^{(n)}}$. Let us now fix $t \in [\hat{T}, \hat{T} + \hat{\Delta}]$. For each $n \in \mathbb{N}$ we choose $k(n) \in (0, \ldots, N_{\Pi^{(n)}})$ such that $|\tau_{k(n)}^{(n)} - t| \leq f(\Pi^{(n)})$ holds. On the one hand this shows $|\tau_{l(n)}^{(n)} - t| \to 0$ for either $l(n) = k(n)$ or $l(n) = k(n) - 1$. On the other hand we defined in 3.11 the piece-wise linear approximative solutions $z^{(n)} \in \text{W}^{1,\infty}([\hat{T}, \hat{T} + \hat{\Delta}], \hat{\mathcal{A}})$ such that they coincide with the discrete solutions at all time points of the corresponding partition. As a consequence we obtain for $l(n) \in \{k(n), k(n) - 1\}$ the estimate

$$\|z_{l(n)}^{(n)} - z(t)\| \leq \|z^{(n)}(\tau_{l(n)}^{(n)}) - z^{(n)}(t)\| + \|z^{(n)}(t) - z(t)\| \to 0 \quad \text{for } n \to \infty,$$

due to the uniform Lipschitz continuity assumed in (3.32) and the convergence of the approximative solutions $z^{(n)}$ to the limit function z which we assumed in (3.33). The Regularity Assumption (3.2) on $\hat{\mathcal{E}}$ and the Assumption (3.34) allow us to pass to the limit in (3.36). This proves (S). ∎

3.6.3 Energy equality

A direct consequence of the stability (S) of the limit function $z \in \text{W}^{1,\infty}([\hat{T}, \hat{T} + \hat{\Delta}], \hat{\mathcal{A}})$ shown in the previous Lemma 3.22 is the following result.

Lemma 3.23 (Lower energy estimate) *Let $z \in W^{1,\infty}([\hat{T}, \hat{T} + \hat{\Delta}], \hat{A})$ satisfy for all $t \in [\hat{T}, \hat{T} + \hat{\Delta}]$ the Stability Condition (S). We assume that there exists a closed set $\mathcal{D} \subset [\hat{T}, \hat{T} + \hat{\Delta}] \times \hat{A}$ such that $\hat{\Psi} \in C(\mathcal{D} \times \mathbb{R}^{3N}, \mathbb{R})$ and for all $t \in [\hat{T}, \hat{T} + \hat{\Delta}]$ we have $(t, z(t)) \in \mathcal{D}$. The functional $\hat{\Psi}(t, z, \cdot)$ shell further satisfy (3.5), i.e. homogeneity of degree one, while $\hat{\mathcal{E}}$ shell satisfy the Regularity Assumption (3.2). Then we find the lower energy estimate*

$$\hat{\mathcal{E}}(t_2, z(t_2)) + \int_{t_1}^{t_2} \hat{\Psi}\big(\tau, z(\tau), \dot{z}(\tau)\big) \; d\tau \geq \hat{\mathcal{E}}(t_1, z(t_1)) + \int_{t_1}^{t_2} \partial_t \hat{\mathcal{E}}(\tau, z(\tau)) \; d\tau$$

for all $t_1, t_2 \in [\hat{T}, \hat{T} + \hat{\Delta}]$ with $t_1 \leq t_2$.

Proof: Since the limit function z satisfies $z \in W^{1,\infty}([\hat{T}, \hat{T} + \hat{\Delta}], \hat{A})$, it is differentiable for almost all $t \in [\hat{T}, \hat{T} + \hat{\Delta}]$. For such a t we have

$$\dot{z}(t) = \lim_{h \to 0} \frac{z(t + h) + z(t)}{h}.$$

We now use the stability (S) of the function z. Choosing $y := z(t + h) \in \hat{A}$ with $h > 0$ we get

$$0 \leq \hat{\mathcal{E}}(t, z(t + h)) - \hat{\mathcal{E}}(t, z(t)) + \hat{\Psi}(t, z(t), z(t + h) - z(t)).$$

Dividing this inequality by $h > 0$, exploiting the continuity and the homogeneity of degree one of $\hat{\Psi}(t, z, \cdot)$ and taking $h \to 0$ we deduce

$$0 \leq D\hat{\mathcal{E}}(t, z(t))\dot{z}(t) + \hat{\Psi}(t, z(t), \dot{z}(t)) \quad \text{for almost all } t \in [\hat{T}, \hat{T} + \hat{\Delta}].$$

Note that the continuity of $\hat{\Psi}$ implies the integrability of $\hat{\Psi}(t, z(t), \dot{z}(t))$, thus integrating from t_1 to $t_2 \in [\hat{T}, \hat{T} + \hat{\Delta}]$ leads to the lower energy estimate

$$\hat{\mathcal{E}}(t_2, z(t_2)) + \int_{t_1}^{t_2} \hat{\Psi}\big(\tau, z(\tau), \dot{z}(\tau)\big) \; \mathrm{d}\tau \geq \hat{\mathcal{E}}(t_1, z(t_1)) + \int_{t_1}^{t_2} \partial_t \hat{\mathcal{E}}(\tau, z(\tau)) \; \mathrm{d}\tau.$$

∎

Before we prove the opposite upper energy estimate we show a technical lemma which will be useful several times in the thesis.

Lemma 3.24 (Lower sequentially semi-continuity) *If the functional $\hat{\Psi}(t, z, \cdot)$ satisfies the homogeneity of degree one (3.5) and the triangle inequality (3.6). For the functions $z, z^{(n)} \in W^{1,\infty}([\hat{T}, \hat{T} + \hat{\Delta}], \hat{A})$ we claim*

$$\|z^{(n)} - z\|_{C^0([\hat{T}, \hat{T} + \hat{\Delta}], \mathbb{R}^{3N})} \to 0, \; (n \to \infty) \quad and \quad \|\dot{z}^{(n)}\|_{L^\infty([\hat{T}, \hat{T} + \hat{\Delta}], \mathbb{R}^{3N})} \leq c_{\text{lip}}. \quad (3.37)$$

Further we assume the existence of a closed set $\mathcal{D} \subset [\hat{T}, \hat{T} + \hat{\Delta}] \times \hat{A}$ such that $\hat{\Psi} \in C(\mathcal{D} \times \mathbb{R}^{3N}, [0, \infty))$ and $(t, z(t)) \in \mathcal{D}$ for all $t \in [\hat{T}, \hat{T} + \hat{\Delta}]$. Then for $\bar{T} = \hat{T} + \hat{\Delta}$ we find

$$\int_{\hat{T}}^{\bar{T}} \hat{\Psi}(t, z(t), \dot{z}(t)) \mathrm{d}t \leq \liminf_{n \to \infty} \int_{\hat{T}}^{\bar{T}} \hat{\Psi}(t, z(t), \dot{z}^{(n)}(t)) \mathrm{d}t.$$

Proof: Due to the assumption (3.37) the sequence $(z^{(n)})_{n \in \mathbb{N}}$ is uniformly bounded in $W^{1,\infty}([\hat{T}, \hat{T} + \hat{\Delta}], \mathbb{R}^{3N})$. Thus, for all $1 \leq p < \infty$ there exists a subsequence, which we still denote by $(z^{(n)})_{n \in \mathbb{N}}$, such that

$$z^{(n)} \rightharpoonup z \quad \text{in } W^{1,p}([\hat{T}, \hat{T} + \hat{\Delta}], \mathbb{R}^{3N}).$$

By a standard result of the direct calculus of variations, see [Dac08] the operator $I(u) := \int_{\hat{T}}^{\bar{T}} \hat{\Psi}(t, z(t), \dot{u}(t)) dt$ is sequentially weakly lower semi-continuous on $W^{1,p}([\hat{T}, \hat{T} + \hat{\Delta}], \mathbb{R}^{3N})$ if the function $f(t, v) := \hat{\Psi}(t, z(t), v)$ is positive, continuous and convex with respect to $v \in \mathbb{R}^{3N}$. The convexity follows from the Assumption (3.5) and (3.6) on $\hat{\Psi}$, see Lemma 3.3, while the continuity follows directly from our assumptions on the set $\mathcal{D}$. ∎

Before we prove the upper energy estimate we recall that the lower energy energy estimate of Lemma 3.23 is equivalent to the monotonicity of the function

$$f(t) := \hat{\mathcal{E}}(t, z(t)) + \int_{\hat{T}}^{t} \hat{\Psi}(\tau, z(\tau), \dot{z}(\tau)) \, d\tau - \int_{\hat{T}}^{t} \partial_t \hat{\mathcal{E}}(\tau, z(\tau))$$

on the interval $[\hat{T}, \hat{T} + \hat{\Delta}]$. In the next lemma we show the opposite estimate but only for the initial and end time, i.e. $f(\bar{T}) \leq f(\hat{T})$ with $\bar{T} := \hat{T} + \hat{\Delta}$. Together with the monotonicity this proves the equality $f(t) = f(\hat{T})$ for all $t \in [\hat{T}, \hat{T} + \hat{\Delta}]$. Recalling the definition of f it is easy to see that this equality is equivalent to the desired energy Equality (E).

Lemma 3.25 (upper energy estimate) *We assume (3.31)–(3.34), the regularity (3.2) of $\hat{\mathcal{E}}$ and homogeneity of degree one (3.5) and the triangle inequality (3.6) for $\hat{\Psi}$ then the limit function z satisfies for $\bar{T} := \hat{T} + \hat{\Delta}$ the upper energy estimate*

$$\hat{\mathcal{E}}(\bar{T}, z(\bar{T})) + \int_{\hat{T}}^{\bar{T}} \hat{\Psi}(\tau, z(\tau), \dot{z}(\tau)) \, d\tau \leq \hat{\mathcal{E}}(\hat{T}, \hat{z}_0) + \int_{\hat{T}}^{\bar{T}} \partial_t \hat{\mathcal{E}}(\tau, z(\tau)) d\tau. \quad (3.38)$$

Proof: We start by deriving a discrete version of (3.38) and consider our sequence of discrete solutions $(z_k^{(n)})_{k=1,\dots,N_{\Pi(n)}}$ which we assumed to exists in Assumption (3.31). Let us fix $n \in \mathbb{N}$ and choose $y = z_{k-1}^{(n)}$ in (IP). Then for any $k = 1, \dots, N_{\Pi(n)}$, we have

$$\hat{\mathcal{E}}(t_k^{(n)}, z_k^{(n)}) + \hat{\Psi}(t_{k-1}^{(n)}, z_{k-1}^{(n)}, z_k^{(n)} - z_{k-1}^{(n)}) \leq \hat{\mathcal{E}}(t_k^{(n)}, z_{k-1}^{(n)}).$$

Here we exploited that the mapping $\hat{\Psi}(t, z, \cdot) : \mathbb{R}^3 \rightarrow [0, \infty)$ is homogeneous of degree one (see Assumption (3.5)) and hence $\hat{\Psi}(t, z, 0) = 0$ holds.

Note that the approximative solution $z^{(n)}$ is defined as a piece-wise interpolation between the values of the discrete solution $(z_k^{(n)})_{k=1,\dots,N_{\Pi(n)}}$ and consequently we have $\dot{z}^{(n)}(t) \equiv \dfrac{z_k^{(n)} - z_{k-1}^{(n)}}{t_k - t_{k-1}}$ for all $t \in (t_{k-1}^{(n)}, t_k^{(n)})$. Again we use the homogeneity of degree one of $\hat{\Psi}(t, z, \cdot)$ to see that the above inequality is equivalent to

$$
\begin{aligned}
&\hat{\mathcal{E}}(t_k^{(n)}, z^{(n)}(t_k^{(n)})) + \int_{t_{k-1}^{(n)}}^{t_k^{(n)}} \hat{\Psi}\big(t_{k-1}^{(n)}, z_{k-1}^{(n)}, \dot{z}^{(n)}(\tau)\big)\, \mathrm{d}\tau \\
&\leq\ \hat{\mathcal{E}}(t_{k-1}^{(n)}, z^{(n)}(t_{k-1}^{(n)})) + \int_{t_{k-1}^{(n)}}^{t_k^{(n)}} \partial_t \hat{\mathcal{E}}(\tau, z_{k-1}^{(n)})\, \mathrm{d}\tau.
\end{aligned}
\tag{3.39}
$$

Next, we want to sum equation (3.39) over k. For this reason we make both integrands independent of the index k and we define the piecewise constant functions $\overline{z}^{(n)}$ and $\overline{t}^{(n)}$ by $\overline{z}^{(n)}(t) := z_k^{(n)}$ and $\overline{t}^{(n)}(t) := t_k^{(n)}$ for all $t \in [t_{k-1}^{(n)}, t_k^{(n)})$ and all $k = 1, \ldots, N_{\Pi^{(n)}}$. Note that the construction is such that we have $\big(\overline{t}^n(\tau), \overline{z}^n(\tau)\big) \in \mathcal{D}$ for all times $\tau \in [\hat{T}, \hat{T} + \hat{\Delta}]$. Hence by adding up (3.39), we obtain for arbitrary $t^{(n)} \in \Pi^{(n)}$

$$
\hat{\mathcal{E}}(\overline{T}, z^{(n)}(\overline{T})) + \int_{\hat{T}}^{\overline{T}} \hat{\Psi}\big(\overline{t}^{(n)}(\tau), \overline{z}^{(n)}(\tau), \dot{z}^{(n)}(\tau)\big)\mathrm{d}\tau \leq \hat{\mathcal{E}}(\hat{T}, \hat{z}_0) + \int_{\hat{T}}^{\overline{T}} \partial_t \hat{\mathcal{E}}\big(\tau, \overline{z}^{(n)}(\tau)\big)\mathrm{d}\tau.
\tag{3.40}
$$

It remains for us to choose a subsequence such that (3.40) converges to (3.38). Due to our convergence Assumption (3.33) for the functions $z^{(n)}$ and our regularity Assumption (3.2) on $\hat{\mathcal{E}}$ we immediately obtain the convergence of the energy terms on the left side of equation (3.40),i.e.

$$
\hat{\mathcal{E}}\big(\overline{T}, z^{(n)}(\overline{T})\big) \to \hat{\mathcal{E}}\big(\overline{T}, z(\overline{T})\big) \quad \text{for } n \to \infty.
$$

The integrands of the integrals on both sides are uniformly bounded due to the uniform boundedness of our approximative solutions and the continuity of the functions $\hat{\Psi}$ and $\partial_t \hat{\mathcal{E}}(t, \cdot)$. By our construction of the piecewise constant functions $\overline{z}^{(n)}$ we derive from the uniform convergence assumed in (3.33) the pointwise convergence $\overline{z}^{(n)}(t) \to z(t)$ for all $t \in [\hat{T}, \hat{T} + \hat{\Delta}]$. This implies the pointwise convergence of the integrands on the right hand side and due to the continuity of $\partial_t \hat{\mathcal{E}}(t, \cdot)$ we obtain

$$
\int_{\hat{T}}^{t} \partial_t \hat{\mathcal{E}}(\tau, \overline{z}^{(n)}(\tau))\mathrm{d}\tau \to \int_{\hat{T}}^{t} \partial_t \hat{\mathcal{E}}(\tau, z(\tau))\mathrm{d}\tau \quad \text{for } n \to \infty.
$$

Next, we want to replace in (3.40) the integrand $\hat{\Psi}\big(\overline{t}^{(n)}(\tau), \overline{z}^{(n)}(\tau), \dot{z}^{(n)}(\tau)\big)$ on the left hand side by $\hat{\Psi}\big(\tau, z(\tau), \dot{z}^{(n)}(\tau)\big)$. Due to the uniform Lipschitz continuity $\|\dot{z}^{(n)}\| \leq \mathrm{c}_{\mathrm{Lip}}$ we have $\big(\overline{t}^{(n)}(\tau), \overline{z}^{(n)}(\tau), \dot{z}^{(n)}(\tau)\big) \in \mathcal{D} \times \mathcal{B}_{\mathrm{c}_{\mathrm{Lip}}}(0)$ for all times $\tau \in [\hat{T}, \hat{T} + \hat{\Delta}]$. But because of (3.34) the functional $\hat{\Psi}$ is uniformly continuous on the bounded set $\mathcal{D} \times \mathcal{B}_{\mathrm{c}_{\mathrm{Lip}}}(0)$. Hence it is sufficient to prove $\|\big(\overline{t}^{(n)}, \overline{z}^{(n)}, \dot{z}^{(n)}\big) - \big(\cdot, z, \dot{z}^{(n)}\big)\|_{\mathrm{C}^0([\hat{T}, \hat{T}+\hat{\Delta}], \mathbb{R}^{6N+1})} \to 0$ but this follows directly from the convergence Assumption (3.33) and $f_{\Pi^{(n)}} \to 0$. All this together shows that the difference of the

integrands $|\bar{\Psi}^{(n)}\big(t,\overline{z}^{(n)}(t),\dot{z}^{(n)}(t)\big) - \hat{\Psi}\big(t,z(t),\dot{z}^{(n)}(t)\big)| \to 0$ vanishes uniformly in
$t \in [\hat{T},\hat{T}+\hat{\Delta}]$ for $n \to \infty$. This finally implies

$$\int_{\hat{T}}^{\bar{T}} \bar{\Psi}\big(\tau,\overline{z}^{(n)}(\tau),\dot{z}^{(n)}(\tau)\big) - \hat{\Psi}\big(\tau,z(\tau),\dot{z}^{(n)}(\tau)\big)\mathrm{d}\tau \to 0 \quad \text{for } n \to \infty.$$

Summarizing the last convergence results and taking the $\liminf_{n\to\infty}$ on both sides
of equation (3.40) we get

$$\hat{\mathcal{E}}(\bar{T},z(\bar{T})) + \liminf_{n\to\infty} \int_{\hat{T}}^{\bar{T}} \hat{\Psi}\big(\tau,z(\tau),\dot{z}^{(n)}(\tau)\big)\mathrm{d}\tau \leq \hat{\mathcal{E}}(\hat{T},\hat{z}_0) + \int_{\hat{T}}^{\bar{T}} \partial_t\hat{\mathcal{E}}(\tau,z(\tau))\mathrm{d}\tau.$$
(3.41)

We now apply the previous Lemma 3.24 and derive the desired upper energy inequality (3.38).   ∎

We now summarize the result (3.35) on the initial condition, the results of Lemma
3.22 on the stability (S) and of Lemma 3.23 and 3.25 on the energy equality (E).

Theorem 3.26 (Existence result) *Let the regularity (3.2) on $\hat{\mathcal{E}}$ hold. For the
dissipation $\hat{\Psi}$ we assume the regularity (3.4) in its weak version and with respect to
its third argument the homogeneity of degree one (3.5) and triangle inequality (3.6).
Further we assume for a given initial value $\hat{z}_0 \in \hat{\mathcal{A}}$ and for a sequence of partitions
$(\Pi^{(n)})_{n\in\mathbb{N}}$ of $[\hat{T},\hat{T}+\hat{\Delta}]$ with $f_{\Pi^{(n)}} \to 0$ the existence of a sequence of discrete solutions
$(z_k^{(n)})_{k=1,\dots,N_{\Pi^{(n)}}}$ (3.31) of the corresponding incremental problems (IP). Let this
sequence be uniformly Lipschitz continuous (3.32) and the corresponding sequence
of approximative solutions $z^{(n)} \in W^{1,\infty}([\hat{T},\hat{T}+\hat{\Delta}],\hat{\mathcal{A}})$ have a convergent subsequence
(3.33). Finally we claim (3.34), i.e. the existence of a closed set $\mathcal{D} \subset [\hat{T},\hat{T}+\hat{\Delta}] \times \hat{\mathcal{A}}$
such that*

$$\hat{\Psi} \in \mathrm{C}\left(\mathcal{D} \times \mathbb{R}^{3N},\mathbb{R}\right) \quad \text{and}$$
$$(t_k,z_k)_{k=1,\dots,N_\Pi} \subset \mathcal{D} \quad \text{for solutions of (IP).}$$

*Then the limit function $z \in W^{1,\infty}([\hat{T},\hat{T}+\hat{\Delta}],\hat{\mathcal{A}})$ provides us with a solution of the
energetic Problem 3.1.*

Corollary 3.27 (Proof of Existence Theorem 3.2) *The generic Assumptions
(3.1)–(3.11) with the Assumption (3.4) in its strong version are sufficient to assure
the existence of a solution $z \in W^{1,\infty}([\hat{T},\hat{T}+\hat{\Delta}],\hat{\mathcal{A}})$ of the energetic Problem 3.1.*

Proof: Using the Assumptions (3.1)–(3.11) with the Assumption (3.4) only in its
weak version we can apply the main result of the convergence section, i.e. Theorem
3.20. This theorem assures us the additional Assumptions (3.31)-(3.33) to hold but
not the last missing Assumption (3.34). This is why we need the strong version of
(3.4) since it directly implies (3.34) with the choice $\mathcal{D} = [\hat{T},\hat{T}+\hat{\Delta}] \times \mathcal{A}$.   ∎

Chapter 4

Making and losing contact (one particle)

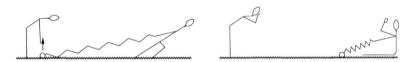

Figure 4.1: Contact vs. non-contact: discontinuous dissipation Ψ

4.1 Overview

In this chapter we incorporate, compared to the previous chapter, the additional difficulty of making and losing contact. This leads to a discontinuous dissipation functional Ψ. But, we still assume the admissible set $\mathcal{A}$ to be convex. We constrain ourself to a single particle situation and assume $\mathcal{A} := \{z \in \mathbb{R}^3 \ : \ z_3 \geq 0\}$ such that we have a flat obstacle boundary $\partial\mathcal{A} = \{z \in \mathbb{R}^3 \ : \ z_3 = 0\}$.

There a two principal ideas in this chapter. The first idea is presented in the Section 4.3. Here we show how we deal with the difficulty of making and losing contact in the one particle situation. We consider two subproblems, one neglecting dissipation and one restricting the solution to the boundary. Consequently both subproblems will have continuous dissipation functionals and we can use the results of the previous Chapter 3 to construct solutions. The surprising result is that one can glue the solutions together to obtain a solution of the full problem. Unfortunately this central trick only works in the one particle situation.

The second principal idea is presented in Section 4.4. Here we revisit the contact subproblem in which we restricted the particle to the boundary. But this time we prove the existence under a weaker assumption on the relation between elasticity and friction, i.e. $\forall v \in \mathcal{T}_{\partial \mathcal{A}}(Z_0), \|v\| = 1$ we assume

$$v^\top H(T_0, Z_0)v + \sigma(T_0, Z_0)\frac{(M(Z_0)v)^\top}{\|M(Z_0)v\|}DM(Z_0)[v, v] + \|M(Z_0)v\|D\sigma(T_0, Z_0)v > 0.$$
(O5)

This improved Assumption (O5) for the constraint problem will enter directly in the existence theorem of the full problem. The special feature of (O5) is that it evaluates the change of frictional and dissipational forces with respect to the same sliding direction v. This is an improvement with respect to the corresponding assumption $\alpha > q$ of Chapter 3. There the minimal change of elastic forces is compared with the maximal change of frictional forces. The difficulty in using this improved Assumption (O5) is that solutions might oscillate and thus change their sliding direction quite fast. Since our main recursive estimate contains naturally two different sliding directions $v_1 = \frac{z_k - z_{k-1}}{\|z_k - z_{k-1}\|}$ and $v_2 = \frac{z_{k-1} - z_{k-2}}{\|z_{k-1} - z_{k-2}\|}$ we run into technical difficulties. The principal idea is first to identify a path or two-dimensional manifold of the set $[0, T] \times \mathcal{A}$ such that high oscillations occur only close to it. Second, we increase artifically the convexity of the energy functional in some neighborhood of this path such that locally we have a strong convexity assumption again whenever the above weak assumption causes difficulties in arguing.

4.2   Assumptions, problem and result

We define our admissible set as

$$\mathcal{A} := \left\{ z \in \mathbb{R}^3 \ : \ z_3 \geq 0 \right\}.$$
(O0)

As a consequence we have $\partial \mathcal{A} = \{ z \in \mathbb{R}^3 \ : \ z_3 \equiv 0 \}$ and for $z \in \partial \mathcal{A}$ we define the normal vector $\nu(z)$ via

$$\nu(z) := \begin{pmatrix} 0 \\ 0 \\ -1 \end{pmatrix}$$

and the tangential plane by

$$\mathcal{T}_{\partial \mathcal{A}}(z) := \left\{ v \in \mathbb{R}^3 \ : \ v_3 \equiv 0 \right\}.$$
(4.1)

Further, for $z \in \mathcal{A}$ we define the set of constraint forces

$$\mathcal{N}_{\mathcal{A}}(z) \subset \mathbb{R}^{3^*}, \quad \mathcal{N}_{\mathcal{A}}(z) := \begin{cases} \left\{ \lambda \nu^\top(z) \ : \ \lambda \geq 0 \right\} & \text{for } z \in \partial \mathcal{A} \\ \{0\} \in \mathbb{R}^{3^*} & \text{for } z \in \text{int}\mathcal{A}. \end{cases}$$
(4.2)

As usual we assume that our energy functional is twice differentiable

$$\mathcal{E} \in C^2\big([0,T] \times \mathcal{A}, [0,\infty)\big). \tag{O1}$$

The next generic assumption is on the regularity of the matrix of friction

$$M \in C^1\big(\partial\mathcal{A}, \mathbb{R}^{3\times3}\big) \text{ with } M(z)\nu(z) = 0 \text{ for all } z \in \partial\mathcal{A} \tag{O2}$$

$$\text{and } \min\big\{\|M(Z_0)v\| \ : \ v \in T_{\partial\mathcal{A}}(Z_0) \cap \mathbb{S}^2\big\} > 0 \text{ if } Z_0 \in \partial\mathcal{A},$$

with $\mathbb{S}^2 \subset \mathbb{R}^3$ begin defined as in (3.3). Note that we exclude here the possibility of $M(Z_0) = 0 \in \mathbb{R}^{3\times3}$ for $Z_0 \in \partial\mathcal{A}$ and especially the possibility of $M \equiv 0$. Of course, this is somehow against our intuition since in the absence of friction the problem should be easier to solve. Hence we exclude the most simple situations. The reason why we do this is less a mathematical one and more due to simplification of notation. If we would replace in the modeling the matrix valued function M by a more complicated product μM_1 consisting of a scalar coefficient of friction $\mu \in C^1(\partial\mathcal{A}, [0,\infty))$ and a standardized matrix valued function $M_1 \in C^1(\partial\mathcal{A}, \mathbb{R}^{3\times3})$ satisfying $\|M_1(z)\| = 1$ for all $z \in \partial\mathcal{A}$ then the right assumption would be

$$\mu \in C^1(\partial\mathcal{A}, [0,\infty)), M_1 \in C^1(\partial\mathcal{A}, \mathbb{R}^{3\times3})$$

$$\text{and } M_1(z)v = 0 \Leftrightarrow v = \pm\nu(z)\|v\| \text{ for all } z \in \partial\mathcal{A}.$$

Note that here $\mu(z) = 0$ is allowed and it is only excluded that there exist a tangential direction causing no friction while another direction does cause friction. Such a singular behavior is physically not reasonable and hence this assumption is no real limitation in modeling. But we do not prove this since we preferred to keep notation simple and to introduce only a single matrix valued function M.

We now introduce the *dissipational metric* $\Psi : [0,T] \times \mathcal{A} \times \mathbb{R}^3 \to [0,\infty)$ by

$$\Psi(t,z,v) := \begin{cases} \sigma(t,z)_+ \|M(z)v\| & \text{if } z \in \partial\mathcal{A} \\ 0 & \text{else} \end{cases}, \tag{4.3}$$

and the normal force $\sigma(t,z) = -D\mathcal{E}(t,z)\nu(z)$ for $z \in \partial\mathcal{A}$ and $\sigma_+ := \max\{0,\sigma\}$. Further $\|\cdot\|$ is the usual Euclidian norm.

After having introduced the dissipational distance we establish our assumption on the initial values $T_0 \in [0,T)$ and $Z_0 \in \mathcal{A}$

$$0 \in D\mathcal{E}(T_0, Z_0) + \partial_v\Psi(T_0, Z_0, 0) + \mathcal{N}_\mathcal{A}(Z_0), \tag{O3}$$

Further, we denote the Hessian matrix of $\mathcal{E}$ with respect to z by $H(t,z) = D^2\mathcal{E}(t,z) \in \mathbb{R}^{3\times3}$. We now assume that there exists a positive constant $\alpha_* > 0$ such that $\mathcal{E}$ is α_*-uniformly convex in its second variable, i.e.

$$v^\top H(T_0, Z_0)v \geq \alpha_*\|v\|^2 \qquad \text{for all } v \in \mathbb{R}^3. \tag{O4}$$

While the above assumptions are somehow classical, the following assumption reveals the nature of our problem and governs the interplay between the different physical quantities. If $Z_0 \in \partial \mathcal{A}$ then we assume for all $v \in \mathcal{T}_{\partial \mathcal{A}}(Z_0)$ with $\|v\| = 1$ the following inequality holds

$$v^\top \mathrm{H}(T_0, Z_0)v + \sigma(T_0, Z_0)\frac{(\mathrm{M}(Z_0)v)^\top}{\|\mathrm{M}(Z_0)v\|}\mathrm{DM}(Z_0)[v, v] + \|\mathrm{M}(Z_0)v\|\mathrm{D}\sigma(T_0, Z_0)v > 0.$$
(O5)

Here we make the notational convention $\big(\mathrm{DM}(z)[u, v]\big)_i = \sum_{j,k=1}^3 \partial_{z_k}\mathrm{M}_{ij}(z)u_j v_k$ for $i = 1, \ldots, 3$ and $u, v \in \mathbb{R}^3$.

Problem 4.1 *For given initial time $T_0 \in [0, T)$ and initial state $Z_0 \in \mathcal{A}$ find a positive time span $\Delta \in (0, T - T_0]$ and a solution $z \in \mathrm{W}^{1,\infty}([T_0, T_0+\Delta], \mathcal{A})$ such that the initial condition $z(T_0) = Z_0$ is satisfied and such that for almost all $t \in [T_0, T_0+\Delta]$ the following differential inclusion holds*

$$0 \in \mathrm{D}\mathcal{E}(t, z(t)) + \partial_v \Psi(t, z(t), \dot{z}(t)) + \mathcal{N}_\mathcal{A}(z(t)) \quad \subset \mathbb{R}^{3^*}.$$
(DI)

Theorem 4.2 (Existence) *Let the Assumptions (O0)–(O3) hold.*

1. **(no contact)** *If we have $Z_0 \in \mathrm{int}\mathcal{A}$ we assume the convexity Assumption (O4) on $\mathcal{E}$,*

2. **(positive normal force)** *if we have $-\mathrm{D}\mathcal{E}(T_0, Z_0)\nu(Z_0) > 0$ (this implies $Z_0 \in \partial\mathcal{A}$ due to (O3)) we assume the joint convexity (O5) of $\mathcal{E}$ and Ψ,*

3. **(contact and no normal force)** *if we have $Z_0 \in \partial\mathcal{A}$ and $\mathrm{D}\mathcal{E}(T_0, Z_0)\nu(Z_0) = 0$ we assume (O4) and (O5),*

then there exists a positive time span $\Delta > 0$ and a solution $z \in \mathrm{W}^{1,\infty}([T_0, T_0+\Delta], \mathcal{A})$ of Problem 4.1.

Proof: We give a first simplified proof of the three cases in the same order in the following three Subsections 4.3.1–4.3.3. We refer for this to the Corollaries 4.5, 4.13 and 4.18 at the end of each of these subsections. The simplification consists in replacing the Joint Convexity Assumption (O5) by a stronger convexity assumption denoted by (O5*). The full proof using the weak Convexity Assumption (O5) is given in Corollary 4.40 in the following Section 4.4. ∎

4.3   Decomposition into two subproblems

4.3.1   Friction-free subproblem

In this section we present a subproblem in which we neglect any friction. For this we introduce the index 'f' for *friction-free* in the notation to distinguish the data or the solution of such a friction-free problem of others. We define the dissipation functional $\Psi_f \equiv 0$. The solution $z_f \in W^{1,\infty}([T_0, T_0 + \Delta_f], \mathcal{A})$ of such a friction-free subproblem will be useful to prove Part 1 and 3 of Theorem 4.2 since in both situations the initial dissipation is zero. In fact, the friction-free solution z_f will represent the solution in Part 1 and will be used to construct a solution in Part 3. We now present the problem.

Problem 4.3 (Friction-free subproblem) *For given initial time $T_0 \in [0, T)$ and initial state $Z_0 \in \mathcal{A}$ find a positive time span $\Delta_f \in (0, T - T_0]$ and a solution $z_f \in W^{1,\infty}([T_0, T_0 + \Delta_f], \mathcal{A})$ such that the initial condition $z_f(T_0) = Z_0$ is satisfied and such that for almost all $t \in [T_0, T_0 + \Delta_f]$ the following differential inclusion holds*

$$0 \in D\mathcal{E}(t, z_f(t)) + \mathcal{N}_{\mathcal{A}}(z_f(t)) \quad \subset \mathbb{R}^{3^*}. \tag{DI$_f$}$$

Our choice $\Psi_f \equiv 0$ avoids the discontinuity with respect to the state z which still appears in the original dissipation Ψ defined in (4.3). Hence, we satisfy one of the crucial assumptions of our Generic Existence Theorem 3.2 which we are going to apply. A further important assumption of Theorem 3.2 is equivalent to the initial local stability

$$0 \in D\mathcal{E}(T_0, Z_0) + \partial_v \Psi_f(T_0, Z_0, 0) + \mathcal{N}_{\mathcal{A}}(Z_0).$$

This is our initial Assumption (O3) but we replaced Ψ by $\Psi_f \equiv 0$ here. Since we want to start with the same initial values T_0, Z_0 and work with the same assumptions as for our original problem we have to restrict ourself to the cases of a friction-free initial situation for the original data, i.e. $\Psi(T_0, Z_0, \cdot) \equiv 0$. We will assure this by assuming additionally $\sigma(T_0, Z_0) = 0$, see (4.3). Hence, in solving the original Problem 4.1 we only consider the auxiliary friction-free problem if the initial normal force is zero.

Theorem 4.4 (Existence for friction-free subproblem) *Assume (O0), (O1), (O3) and (O4). Further we assure our initial situation to be friction-free by assuming $\sigma(T_0, Z_0) = 0$.*
Then there exists a time span $\Delta_f > 0$ and a unique solution $z_f \in W^{1,\infty}([T_0, T_0 + \Delta_f], \mathcal{A})$ of the friction-free subproblem 4.3.

Further, for each $\rho > 0$ there exists a time span $\Delta_\rho \in (0, \Delta_f]$ such that for almost all $t \in [T_0, T_0 + \Delta_\rho]$ the estimate

$$\|\dot{z}_f(t)\| \leq \frac{\|\partial_t D\mathcal{E}(T_0, Z_0)\|}{\alpha_*} + \rho \quad \text{holds.} \tag{4.4}$$

Proof: We start by extending the Convexity Assumption (O4) to some neighborhood of (T_0, Z_0). For given radius $r > 0$ and given time span $\hat{\Delta} > 0$ we introduce the set

$$\hat{\mathcal{A}} = \mathcal{A} \cap \mathcal{B}_r(Z_0) \quad \subset \mathbb{R}^3 \quad \text{and the cylinder}$$
$$\hat{\mathcal{C}} = [T_0, T_0 + \hat{\Delta}] \times \hat{\mathcal{A}}.$$

We specify these neighborhoods. For $\rho > 0$ we choose, due to the regularity (O1) of $\mathcal{E}$ and its convexity (O4), the parameters $r, \hat{\Delta} > 0$ such that the constants

$$\hat{c}_1 = \max\left\{ \|\partial_t D\mathcal{E}(t, z)\| \; : \; (t, z)\| \in \hat{\mathcal{C}} \right\} \quad \text{and}$$
$$\hat{\alpha}_* = \min\left\{ v^\top H(t, z)v \; : \; (t, z) \in \hat{\mathcal{C}}, v \in \mathbb{S}^2 \right\}$$

satisfy $\hat{\alpha}_* > 0$ and

$$\frac{\hat{c}_1}{\hat{\alpha}_*} \leq \frac{\|\partial_t D\mathcal{E}(T_0, Z_0)\|}{\alpha_*} + \rho.$$

Due to the resulting $\hat{\alpha}_*$-uniform convexity of $\mathcal{E}(t, \cdot)$ we either define z_f for $t \in [T_0, T_0 + \hat{\Delta}]$ as the unique minimizer

$$z_f(t) := \operatorname{argmin}\left\{ \mathcal{E}(t, y) \; : \; y \in \hat{\mathcal{A}} \right\} \tag{4.5}$$

and check the initial condition $z_f(T_0) = Z_0$ via the initial force balance (O3) and the additional assumption $\sigma(T_0, Z_0) = 0$ or $\Psi(T_0, Z_0, \cdot) \equiv 0$ respectively. Both assumptions together imply

$$0 \in D\mathcal{E}(T_0, Z_0) + \mathcal{N}_{\hat{\mathcal{A}}}(Z_0). \tag{4.6}$$

Or, we construct z_f using Theorem 3.2 with the choice $\hat{\Psi} \equiv \Psi_f \equiv 0$. After verifying the Assumptions (3.1)–(3.11) with the choice $\hat{c}_2 = \hat{q} = 0$ we find a function $z_f \in W^{1,\infty}([T_0, T_0 + \hat{\Delta}], \hat{\mathcal{A}})$ that satisfies for all $t \in [T_0, T_0 + \hat{\Delta}]$ the Stability Condition (S) with $\hat{\Psi} \equiv 0$, i.e.

$$\mathcal{E}(t, z_f(t)) \leq \mathcal{E}(t, y) \quad \text{for all } y \in \hat{\mathcal{A}}.$$

Considering (4.5) we find that both constructions of z_f are equivalent.

In the second part of the proof we now make use of Theorem 2.9 that shows the equivalence of the formulations (S)&(E) and (DI$_f$), i.e. for almost all $t \in [T_0, T_0 + \hat{\Delta}]$ we have

$$0 \in D\mathcal{E}(t, z(t)) + \mathcal{N}_{\hat{\mathcal{A}}}(z(t)).$$

This is almost our desired differential inclusion ($\mathrm{DI_f}$) apart of the generic set $\hat{\mathcal{A}} = \mathcal{A} \cap \mathcal{B}_r(Z_0)$. But since our solution z starts at Z_0 and is Lipschitz continuous it is sufficient to shorten our time interval and to replace $\hat{\Delta}$ by some $\Delta_f \in (0, \hat{\Delta}]$ such that $\|Z_0 - z(t)\| < r$ for all $t \in [T_0, T_0 + \Delta_f]$. Doing this we find $\mathcal{N}_{\hat{\mathcal{A}}}(z(t)) = \mathcal{N}_{\mathcal{A}}(z(t))$ for all $t \in [T_0, T_0 + \Delta_f]$. ∎

Corollary 4.5 (Proof of Theorem 4.2 Part 1, no contact) *The above result proves directly Theorem 4.2, Part 1 no contact.*

Proof: The assumptions $Z_0 \in \mathrm{int}\mathcal{A}$ and (O3) imply the above additional assumption $\sigma(T_0, Z_0) = 0$. Further, there exists $\Delta \in (0, \Delta_f]$ such that $z_f(t) \in \mathrm{int}\mathcal{A}$ and consequently $\Psi(t, z_f(t), \cdot) \equiv 0$ hold for all $t \in [T_0, T_0 + \Delta)$. Hence ($\mathrm{DI_f}$) and (DI) coincide on $[T_0, T_0 + \Delta]$ and we define $z \in \mathrm{W}^{1,\infty}([T_0, T_0 + \Delta], \mathcal{A})$ by restricting z_f on $[T_0, T_0 + \Delta]$. ∎

The following lemma will be the only uniqueness result that we are going to present in this work. It applies to the full Problem 4.1..

Lemma 4.6 (Uniqueness up to contact) *Let us assume that there exists $R > 0$ such that $\mathcal{A} \cap \mathcal{B}_R(Z_0) \subset \mathbb{R}^3$ is convex and the function $\mathcal{E}(t, \cdot)$ is strictly convex on $\mathcal{A} \cap \mathcal{B}_R(Z_0)$ for all $t \in [T_0, T_0 + \Delta]$. Further we assume the function $z \in \mathrm{W}^{1,\infty}([T_0, T_0 + \Delta], \mathcal{A})$ to be a solution of Problem 4.1 with $z(t) \in \mathrm{int}\, (\mathcal{A}) \cap \mathcal{B}_R(Z_0)$ for all $t \in [T_0, T_0 + \Delta)$. Then, the solution is unique on $[T_0, T_0 + \Delta]$.*

Proof: Due to our assumption $z(t) \in \mathrm{int}\mathcal{A}$ we have the equalities $\Psi(t, z(t), \cdot) \equiv 0$ and $\mathcal{N}_{\mathcal{A}}(z(t)) = 0 \in \mathbb{R}^{3^*}$ for all $t \in [T_0, T_0 + \Delta)$ and hence z satisfies

$$\mathrm{D}\mathcal{E}(t, z(t)) \equiv 0 \in \mathbb{R}^3 \qquad (4.7)$$

for all $t \in [T_0, T_0 + \Delta]$. Due to the convexity of $\mathcal{E}(t, \cdot)$ and of the domain there exists at most one function satisfying (4.7). Let us assume that we have any other solution $\hat{z} \in \mathrm{W}^{1,\infty}([T_0, T_0 + \Delta], \mathcal{A})$ of Problem 4.1 starting in the same initial state Z_0. The time $t_* := \max \{t \in [T_0, T_0 + \Delta] : \hat{z}(\tau) = z(\tau) \text{ for all } \tau \in [T_0, t]\}$ defines the moment of separation. If $t_* = T_0 + \Delta$ then the solutions coincide. Hence we assume $t_* \in [T_0, T_0 + \Delta)$ and find $\hat{z}(t_*) \in \mathrm{int}\, (\mathcal{A} \cap \mathcal{B}_R(Z_0))$. Since $\hat{z}$ is continuous there exists a time span $\delta > 0$ such that $\hat{z}(t) \in \mathrm{int}\, (\mathcal{A} \cap \mathcal{B}_R(Z_0))$ holds for all $t \in [t_*, t_* + \delta]$. But this implies $\mathrm{D}\mathcal{E}(t, \hat{z}(t)) \equiv 0$ as mentioned above and thus $\hat{z}(t) = z(t)$ for all $t \in [t_*, t_* + \delta]$. This is a contradiction to the definition of t_*. ∎

Ballard [Bal99] has shown that in case of contact no uniqueness can be expected even for arbitrary small coefficients of friction, i.e. in a situation where our assumptions hold. The following example is easier but violates the Assumption (O5). Anyhow it shows that the above uniqueness result can not be extended any further.

Example 4.7 (Non-uniqueness) *We consider a two-dimensional problem with the admissible set $\mathcal{A} := \{z \in \mathbb{R}^2 : z_2 \geq 0\}$. For this we assume $a > b \geq 0$ and define the energy functional $\mathcal{E}(t, z) := \frac{1}{2} \begin{pmatrix} z_1 & z_2 \end{pmatrix} \begin{pmatrix} a & -b \\ -b & a \end{pmatrix} \begin{pmatrix} z_1 \\ z_2 \end{pmatrix} - \begin{pmatrix} at & -bt \end{pmatrix} \begin{pmatrix} z_1 \\ z_2 \end{pmatrix}$. As a consequence we find*

$$\mathrm{D}\mathcal{E}(t, z) = \begin{pmatrix} z_1 & z_2 \end{pmatrix} \begin{pmatrix} a & -b \\ -b & a \end{pmatrix} - \begin{pmatrix} a & -b \end{pmatrix} t \qquad \in \mathbb{R}^{2^*}.$$

We directly see that $z_{\mathrm{f}}(t) = \begin{pmatrix} t & 0 \end{pmatrix}^\top$ is a solution of Problem 4.1 since $\mathrm{D}\mathcal{E}(t, z_{\mathrm{f}}(t)) \equiv 0$ and thus $\partial_v \Psi(t, z_{\mathrm{f}}(t), \cdot) \equiv 0$ hold for all $t \geq 0$. We assume isotropic friction with a constant coefficient of friction satisfying $\mu \geq \frac{a}{b}$. This contradicts our Assumption (O5). Then we find the function $z_{\mathrm{c}}(t) \equiv 0 \in \mathbb{R}^2$ also to be a solution since

$$-\partial_{z_1}\mathcal{E}(t, z_{\mathrm{c}}(t)) = at \in \mu bt[-1, 1] = \partial_{v_1}\Psi(t, 0, 0) = \partial_{v_1}\Psi(t, z_{\mathrm{c}}(t), \dot{z}_{\mathrm{c}}(t)).$$

4.3.2 Contact subproblem (under strong convexity assumption)

The contact subproblem is the counterpart of the preceding friction-free subproblem. This time we avoid the discontinuity of the dissipation functional Ψ defined in (4.3) by restricting the solution to the boundary $\partial\mathcal{A} = \{z \in \mathbb{R}^3 : z_3 = 0\}$. Hence, the problem is two-dimensional only. Further, we only consider the contact subproblem if the original initial state Z_0 is already in contact, i.e. $Z_0 \in \partial\mathcal{A}$. This assumption is satisfied in Theorem 4.2 Part 2 and 3. The contact solution z_{c} will represent a solution in the situation of Part 2 while it will be used together with the friction-free solution z_{f} to construct a solution for the original Problem 4.1 in the situation of Part 3.

Problem 4.8 (Contact subproblem) *For given initial time $T_0 \in [0, T)$ and initial state $Z_0 \in \partial\mathcal{A}$ find a positive time span $\Delta_{\mathrm{c}} \in (0, T - T_0]$ and a solution $z_{\mathrm{c}} \in \mathrm{W}^{1,\infty}([T_0, T_0 + \Delta_{\mathrm{c}}], \partial\mathcal{A})$ such that the initial condition $z_{\mathrm{c}}(T_0) = Z_0$ is satisfied and such that for almost all $t \in [T_0, T_0 + \Delta_{\mathrm{c}}]$ (for all $t \in [T_0, T_0 + \Delta_{\mathrm{c}}]$ such that $\dot{z}_{\mathrm{c}}(t)$ exists) the following differential inclusion holds*

$$0 \in \mathrm{D}\mathcal{E}(t, z_{\mathrm{c}}(t)) + \partial_v\Psi(t, z_{\mathrm{c}}(t), \dot{z}_{\mathrm{c}}(t)) + \mathcal{N} \quad \subset \mathbb{R}^{3^*} \qquad \text{(DI}_{\mathrm{c}})$$

with the set $\mathcal{N} := \begin{pmatrix} 0 & 0 & \mathbb{R} \end{pmatrix}^\top$.

From a physical point of view the above choice for the set of constraint forces $\mathcal{N}$ expresses that the solution z_{c} is restricted to the boundary $\partial\mathcal{A} = \{z \in \mathbb{R}^3 : z_3 \equiv 0\}$. This also reveals the two-dimensional nature of the problem. We introduce now a new notation and new functionals to be consistent with the setting of the basic mathematical strategy in Chapter 3. In the following a bar ' ¯ ' in the notation indicates that the data is expressed with respect to the plane $\mathbb{R}^2$. We introduce the functionals $\bar{\mathcal{E}} \in C^2([0,T] \times \mathbb{R}^2, \mathbb{R})$ and $\bar{\Psi} \in C([0,T] \times \mathbb{R}^2 \times \mathbb{R}^2, [0,\infty))$ and define them for $t \in [0,T], \bar{z} \in \mathbb{R}^2$ and $\bar{v} \in \mathbb{R}^2$ via

$$\bar{\mathcal{E}}(t,\bar{z}) := \left(t, \begin{pmatrix} \bar{z}_1 & \bar{z}_2 & 0 \end{pmatrix}^\top\right) \quad \text{and} \tag{4.8}$$

$$\bar{\Psi}(t,\bar{z},\bar{v}) := \sigma\left(t, \begin{pmatrix} \bar{z}_1 & \bar{z}_2 & 0 \end{pmatrix}^\top\right)_+ \left\| M\left(\begin{pmatrix} \bar{z}_1 & \bar{z}_2 & 0 \end{pmatrix}^\top\right) \begin{pmatrix} \bar{v}_1 & \bar{v}_2 & 0 \end{pmatrix}^\top \right\|. \tag{4.9}$$

We now present an equivalent problem formulation of the contact subproblem.

Problem 4.9 (Two-dimensional contact subproblem) *For a given time* $T_0 \in [0,T)$ *and initial state* $\bar{z}_0 \in \mathbb{R}^2$ *find a positive time span* $\bar{\Delta} \in (0, T - T_0]$ *and a solution* $\bar{z} \in W^{1,\infty}\left([T_0, T_0 + \bar{\Delta}], \mathbb{R}^2\right)$ *such that the initial condition* $\bar{z}(T_0) = \bar{z}_0$ *is satisfied and such that for almost all* $t \in [T_0, T_0 + \bar{\Delta}]$ *the following differential inclusion holds*

$$0 \in D\bar{\mathcal{E}}(t, \bar{z}(t)) + \partial_{\bar{v}}\bar{\Psi}(t, \bar{z}(t), \dot{\bar{z}}(t)) \quad \subset \mathbb{R}^{2^*}. \tag{$\overline{\text{DI}}$}$$

Lemma 4.10 (Equivalence of contact subproblems) *Let us assume* (O1) *and* (O2). *If the relation* $\bar{z}_0 = \begin{pmatrix} Z_{01} & Z_{02} \end{pmatrix}^\top$ *holds for the initial states, then the Problems 4.8 and 4.9 are equivalent.*

Proof: We make the ansatz $z_{c1} \equiv \bar{z}_1$ and $z_{c2} \equiv \bar{z}_2$. If we further denote by $\partial_{v_1,v_2}\Psi(t,z,v) \subset \mathbb{R}^{2^*}$ the subdifferential of Ψ with respect to the first two components of the third variable, then we find on the one hand the general relation

$$\partial_v \Psi(t,z,v) = \left(\partial_{v_1,v_2}\Psi(t,z,v) \quad 0\right) \subset \mathbb{R}^{3^*}.$$

This formula always holds and is a consequence of $M(z)\nu(z) = 0$, see Assumption (O2). Hence, the third column in (DI_c) reads $-\sigma(t, z_c(t)) \in \mathbb{R}$ and we drop it. On the other hand we find

$$\partial_{v_1,v_2}\Psi(t, z_c(t), v) = \partial_{\bar{v}}\bar{\Psi}(t, \bar{z}(t), \bar{v}).$$

This shows that the first two lines in (DI_c) and $(\overline{\text{DI}})$ coincide for the above ansatz and we establish our result. ∎

In this subsection we will only prove a weaker version of Theorem 4.2 and replace the weak Assumption (O5) by a stronger assumption. For this we introduce the functions c, q and $\alpha_c : [0,T] \times \partial\mathcal{A} \to \mathbb{R}$ via

$$c(t,z) := \|M(z)\| \, |\partial_t \partial_{z_3}\mathcal{E}(t,z)| + \frac{1}{2}\| \left(\partial_t\partial_{z_1}\mathcal{E}(t,z) \quad \partial_t\partial_{z_2}\mathcal{E}(t,z)\right) \|, \tag{4.10}$$

$$q(t,z) := \left(\|DM(z)\|\sigma(t,z)_+ + \|M(z)\| \, \|\left(H_{31}(t,z) \quad H_{32}(t,z)\right)\|\right) \quad \text{and} \tag{4.11}$$

$$\alpha_c(t,z) := \min\left\{v^\top H(t,z)v \, : \, v \in \mathbb{R}^3, v_3 = 0, \|v\| = 1\right\}. \tag{4.12}$$

All norms above are operator norms, for example $\|M(z)\| = \sup\{\|M(z)v\| : v \in \mathbb{S}^2\}$ or $\|DM(z)\| := \sup\{\|DM(z)[v_1, v_2]\| : v_j \in \mathbb{S}^2 \text{ for } j = 1, 2\}$ etc. Instead of (O5) we will claim in the following the stronger assumption

$$q(T_0, Z_0) < \alpha_c(T_0, Z_0). \tag{O5*}$$

Theorem 4.11 (Existence for contact subproblem using (O5*)) *Let the initial value satisfy $Z_0 \in \partial\mathcal{A}$. Further we assume (O0)–(O3) and (O5*). Then there exists a positive time span $\Delta_c > 0$ and a solution $z_c \in W^{1,\infty}([T_0, T_0 + \Delta_c], \partial\mathcal{A})$ of the Contact Subproblem 4.8.*
Further, for each $\rho > 0$ we can choose the time span $\Delta_c > 0$ such that for almost all $t \in [T_0, T_0 + \Delta_\rho]$ we have the estimate

$$\|\dot{z}_c(t)\| \leq \frac{c(T_0, Z_0)}{\alpha_c(T_0, Z_0) - q(T_0, Z_0)} + \rho \tag{4.13}$$

with the functions c, q and α_c as defined in (4.10) – (4.12).

Proof: We will prove existence for the equivalent two-dimensional Problem 4.9. The proof consists of two parts. In the first part we apply the Generic Existence Theorem 3.2 to establish the existence of a solution for an Energetic Problem 3.1. In the second part we use Theorem 2.9 and show that this energetic problem is equivalent to 4.9.
In the first part we have to check the Assumptions (3.1)–(3.11) of Theorem 3.2. We introduce a radius r and a time span $\hat{\Delta}$. For the time being we only assume $r > 0$ and $\hat{\Delta} \in (0, T - T_0]$ and we will specify this choice later on. With the help of $r > 0$ we fix

$$\hat{\mathcal{A}} := \mathcal{B}_r(\bar{z}_0) \subset \mathbb{R}^2.$$

This set $\hat{\mathcal{A}}$ fulfills the Assumption (3.1). We postpone the proof of Assumption (3.3). The Assumptions (3.2)–(3.6) follow directly from the definition of $\bar{\mathcal{E}}$ and $\bar{\Psi}$ and the regularity Assumptions (O1) and (O2). Note that we have $\bar{\Psi} \in C([0, T] \times \mathbb{R}^2 \times \mathbb{R}^2, \mathbb{R})$ and thus the strong version of Assumption (3.4) holds. We still have to assure the inequalities (3.3) and (3.7)–(3.11). For this we introduce the constants $\hat{\alpha}_*, \hat{c}$ and $\hat{q}$ which all depend on the two parameters $r > 0$ and $\hat{\Delta} \in (0, T - T_0]$ and are defined with respect to the original data. This is why we introduce the set $\mathcal{A}_c := \partial\mathcal{A} \cap \mathcal{B}_r(Z_0) \subset \mathbb{R}^3$ which represents the three-dimensional counterpart of the set $\hat{\mathcal{A}} \subset \mathbb{R}^2$ defined above. Further we denote by $\mathcal{C}$ the cylinder $\mathcal{C} := [T_0, T_0 + \hat{\Delta}] \times \mathcal{A}_c$ and we now fix

$$\hat{c}_1 := \frac{1}{2}\|(\partial_t\partial_{z_1}\mathcal{E} \quad \partial_t\partial_{z_2}\mathcal{E})\|_{L^\infty(\mathcal{C}, \mathbb{R}^{2*})}, \qquad \hat{c}_2 := \|M\|_{L^\infty(\mathcal{A}_c, \mathbb{R}^{3\times3})}\|\partial_t\partial_{z_3}\mathcal{E}\|_{L^\infty(\mathcal{C}, \mathbb{R}^*)},$$

$$\hat{c} := \hat{c}_1 + \hat{c}_2,$$

$$\hat{q} := \|DM\|_{L^\infty(\mathcal{A}_c, \mathbb{R}^{3\times3\times3})}\|\sigma_+\|_{L^\infty(\mathcal{C}, \mathbb{R})} + \|M\|_{L^\infty(\mathcal{A}_c, \mathbb{R}^{3\times3})}\|(H_{31}, H_{32})\|_{L^\infty(\mathcal{C}, \mathbb{R}^{2*})} \quad \text{and}$$

$$\hat{\alpha}_* := \min\{v^\top H(t, z)v : v \in \mathbb{S}^2, v_3 = 0, (t, z) \in \mathcal{C}\}.$$

The choice of the constants $\hat{c}_1, \hat{c}_2$ and $\hat{q}$ is such that the Assumptions (3.7)–(3.9) hold. Further all constants depend continuously on r and $\hat{\Delta}$ due to the Assumptions (O1) and (O2) on the original data $\mathcal{E}$ and M. The constants $\hat{c}$ and $\hat{q}$ increase if either r or $\hat{\Delta}$ increase. The opposite holds for $\hat{\alpha}_*$. Due to continuity of the data, we can extend the estimate (O5*) to some neighborhood of $(T_0, \bar{z}_0)$. For any given $\rho > 0$ we find a radius $r > 0$ and a time span $\hat{\Delta} > 0$ such that $\hat{\alpha}_* > \hat{q}$ and additionally we have

$$\frac{\hat{c}}{\hat{\alpha}_* - \hat{q}} \leq \frac{c(T_0, \bar{z}_0)}{\alpha_c(T_0, \bar{z}_0) - q(T_0, \bar{z}_0)} + \rho. \tag{4.14}$$

The estimate $\hat{\alpha}_* > \hat{q}$ matches Assumption (3.10) and is the extended version of Assumption (O5*) that only considers the initial values. Further it implies the $\hat{\alpha}_*$-uniform convexity of $\bar{\mathcal{E}}(t, \cdot)$ on $\hat{\mathcal{A}}$ and thus Assumption (3.3). The last missing assumption to verify is the initial stability (3.11). But due to the convexity of $\bar{\mathcal{E}}(t, \cdot)$ we apply Lemma 2.8 which shows that our initial local stability (O3) is equivalent to (3.11).

We have now checked all the Assumptions (3.1)–(3.11) of the Generic Existence Theorem 3.2. This assures us the existence of a function $\bar{z} \in W^{1,\infty}([T_0, T_0 + \hat{\Delta}], \hat{\mathcal{A}})$ that satisfies the Problem 3.1. Further, it assures as

$$\|\dot{\bar{z}}(t)\| \leq \frac{\hat{c}}{\hat{\alpha}_* - \hat{q}} \quad \text{for almost all } t \in [T_0, T_0 + \hat{\Delta}].$$

Together with (4.14) this proves (4.13).

In the second part of the proof we use again that $\bar{\mathcal{E}}(t, \cdot)$ is convex on $\hat{\mathcal{A}}$ for all $t \in [T_0, T_0 + \hat{\Delta}]$. This allows us to apply Theorem 2.9. which shows that the energetic Problem 3.1 and the following differential inclusion are equivalent. Hence, for almost all $t \in [T_0, T_0 + \hat{\Delta}]$ our solution $\bar{z}$ satisfies

$$0 \in D\bar{\mathcal{E}}(t, \bar{z}(t)) + \partial_{\dot{v}}\bar{\Psi}(t, \bar{z}(t), \dot{\bar{z}}(t)) + \mathcal{N}_{\hat{\mathcal{A}}}(\bar{z}(t)) \quad \subset \mathbb{R}^{2^*}.$$

Apart of the last term this is exactly our desired two-dimensional differential inclusion $(\overline{\text{DI}})$. We recall the definition $\hat{\mathcal{A}} = \mathcal{B}_r(\bar{z}_0) \subset \mathbb{R}^2$. Due to the Lipschitz continuity of $\bar{z}$ and the initial condition $\bar{z}(T_0) = \bar{z}_0$ we can choose $\bar{\Delta} \in (0, \hat{\Delta}]$ such that $\|\bar{z}(t) - \bar{z}_0\| < r$ holds for all $t \in [T_0, T_0 + \bar{\Delta}]$. If we restrict our solution $\bar{z}$ to this interval we find $\mathcal{N}_{\hat{\mathcal{A}}}(\bar{z}(t)) = 0 \in \mathbb{R}^{2^*}$ for all $t \in [T_0, T_0 + \bar{\Delta}]$. This finally proves that $\bar{z} \in W^{1,\infty}([T_0, T_0 + \bar{\Delta}], \mathbb{R}^2)$ is a solution of the two-dimensional Contact Subproblem 4.9 which is equivalent to the Contact Subproblem 4.8. ∎

Lemma 4.12 *Assume* (O0), (O1), $\sigma(T_0, Z_0) > 0$ *and let* $z_c \in W^{1,\infty}([T_0, T_0 + \Delta_c], \partial\mathcal{A})$ *be a solution of the Contact Subproblem 4.8. Then there exists a solution of the original Problem 4.1.*

Proof: If we have a strictly positive initial normal force $\sigma(T_0, Z_0) > 0$ then due to continuity there exists $\Delta \in (0, \Delta_c]$ such that $\sigma(t, z_c(t)) \geq 0$ (or equivalently $-\sigma(t, z_c(t)) \in (-\infty, 0])$ holds for all $t \in [T_0, T_0 + \Delta]$. Inserting this into (DI$_c$) shows that we can replace $\mathcal{N} = \begin{pmatrix} 0 & 0 & \mathbb{R}^* \end{pmatrix}$ by $\mathcal{N}_{\mathcal{A}}(z_c(t)) = \begin{pmatrix} 0 & 0 & (-\infty, 0] \end{pmatrix}$ for all those $t \in [T_0, T_0 + \Delta]$ and we find (DI). This is the proof for the second part of Theorem 4.2. ∎

Corollary 4.13 (Proof of Theorem 4.2 Part 2, using (O5*)**)** *The above theorem and preceding lemma prove the Existence Theorem 4.2 in the case of a positive normal force $\sigma(T_0, Z_0) > 0$ and under the strong Assumption* (O5*).

4.3.3 Merging the two solutions

Why is it not possible to solve Problem 4.1 by a case study as follows: if we start on the boundary (i.e. $Z_0 \in \partial \mathcal{A}$) we take z_c as solution and if we start with no friction (i.e. $\sigma(T_0, Z_0) = 0$) we choose z_f as solution!? The following example shows that this idea is too simple since we have to distinguish a third case, i.e. we start with $Z_0 \in \partial \mathcal{A}$ and $\sigma(T_0, Z_0) = 0$. In this situation we find that within arbitrary short time intervals a solution of Problem 4.1 might switch infinitely many times between having contact with friction and being friction-free and neither z_c nor z_f is the right solution on this interval.

Example 4.14 (Infinitely many switches between z_c **and** z_f**)** *We consider a single particle with two degrees of freedom only and thus define the admissible set via $\mathcal{A} := \{z \in \mathbb{R}^2 : z_2 \geq 0\}$. The energy functional is defined via $\mathcal{E}(t, z) := \sum_{j=1}^2 z_j^2 - f_j(t) z_j$. The tangential external force has the simple form $f_1(t) := t$ while the normal external force is highly oscillating for small t and we define $f_2(0) := 0$ and $f_2(t) := t^2 \sin(1/t)$ for $t > 0$. The coefficient of friction is chosen constant with $\mu = 1/2$ and for the initial values we take $T_0 = 0$ and $Z_0 = 0 \in \mathbb{R}^2$. Our problem to solve is $0 \in \mathrm{D}\mathcal{E}(t, z(t)) + \partial_v \Psi(t, z(t), \dot{z}(t)) + \mathcal{N}_{\mathcal{A}}(z(t)) \subset \mathbb{R}^{2^*}$. In case of having no contact, i.e. $z_2(t) > 0$ this simplifies to*

$$0 = z_1(t) - f_1(t)$$
$$0 = z_2(t) - f_2(t)$$

while in case of contact, i.e. $z_2(t) = 0$ we have to solve

$$0 \in z_1(t) - f_1(t) + \mu\big(-f_2(t)\big)_+ \mathrm{Sign}(\dot{z}_1(t)),$$
$$0 \in z_2(t) - f_2(t) + (-\infty, 0].$$

The solution finally reads for $t > 0$ and $\mu = \frac{1}{2}$

$$z_1(t) = t - \mu t^2 \left(-\sin(\frac{1}{t}) \right)_+ ,$$

$$z_2(t) = t^2 \left(\sin(\frac{1}{t}) \right)_+ .$$

Of course no restriction on $\mu > 0$ is needed to assure existence. since the Hessian

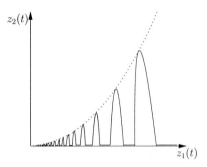

Figure 4.2: Example for infinitely many switches between contact and non-contact

matrix $H = D^2 \mathcal{E}$ *has no off-diagonal entries and thus $\alpha > q = \mu |H_{12}| \equiv 0$ for all $\mu \geq 0$ holds. We have chosen $\mu = \frac{1}{2}$ because this simplifies the representation of the solution. Note that for $\mu = \frac{1}{2}$ we have $\dot{z}_1(t) > 0$ for all $t > 0$ and thus no sticking occurs. Sticking would make the formula for the solution nasty. The trajectory of the solution is presented in Figure 4.2.*

The above example has shown that within arbitrary small time intervals the solution of our Problem 4.1 can switch infinitely many times between $z(t) = z_c(t)$ and $z(t) = z_f(t)$. In the following definition we make a suggestion when to chose $z_c(t)$ and when $z_f(t)$.

Definition 4.15 (Solution candidate) *For the values $Z_0 \in \partial \mathcal{A}, T_0 \geq 0, \Delta > 0$ with $[T_0, T_0+\Delta] \subset [0, T]$ and for solutions $z_f \in W^{1,\infty}([T_0, T_0+\Delta], \mathcal{A})$ and $z_c \in W^{1,\infty}([T_0, T_0+\Delta], \partial \mathcal{A})$ of the Friction-free and Contact Subproblems 4.3 and 4.8, respectively we define the* solution candidate *$z : [T_0, T_0+\Delta] \to \mathcal{A}$ via*

$$z(t) := \begin{cases} z_c(t) & \text{if } \sigma(t, z_c(t)) > 0, \\ z_f(t) & \text{else.} \end{cases} \tag{4.15}$$

Hence, we use both solutions to construct a solution of Problem 4.1. We decide to take the contact solution z_c whenever it is pressed onto the obstacle, i.e.

$\sigma(t, z_c(t)) > 0$. Of course the contact solution $z_c \in W^{1,\infty}([T_0, T_0+\Delta], \partial\mathcal{A})$ is not physically meaningful if it is pulled away from the obstacle but remains due to its construction on the boundary $\partial\mathcal{A}$. But in this situation of negative normal forces (i.e. $\sigma(t, z_c(t)) < 0$ and $\Psi(t, z_c(t), \cdot) \equiv 0$ hold) we decide to define the solution candidate as the friction-free solution z_f. Of course we will have to discuss whether the fact that z_c faces no friction automatically implies the same (no positive normal force) for the frictionless solution z_f. Because only then z_f is physically the right solution. This question will be addressed in Theorem 4.17 below. But before we have to consider the case of $\sigma(t, z_c(t)) = 0$ and ask if the two solutions which might have such a different past match in such a moment? The answer gives the following lemma.

Lemma 4.16 (Lipschitz continuity of z) *Let* (O0) *and* (O1) *hold. For given* $Z_0 \in \partial\mathcal{A}$ *let* $z : [T_0, T_0+\Delta] \to \mathcal{A}$ *be a solution candidate. Further we assume that there exists some* $r > 0$ *such that for all* $t \in [T_0, T_0+\Delta]$ *we have* $z(t) \in \mathcal{A} \cap \mathcal{B}_r(Z_0)$ *and* $\mathcal{E}(t, \cdot)$ *is strictly convex on* $\mathcal{A} \cap \mathcal{B}_r(Z_0)$, *then the function z is Lipschitz continuous, i.e we have*

$$z \in W^{1,\infty}([T_0, T_0+\Delta], \mathcal{A}).$$

Proof: Since both functions z_c and z_f are Lipschitz continuous it is sufficient to prove the continuity of z. The set $\mathcal{P} := \{\tau \in [T_0, T_0+\Delta] : \sigma(\tau, z_c(\tau)) > 0\}$ is open due to the continuity of the normal force σ, see Assumption (O1), and the continuity of the function z_c. Since z coincides with z_c on $\mathcal{P}$ we deduce that it is continuous on $\mathcal{P}$. The same argument holds for all $t \in [T_0, T_0+\Delta]\backslash\text{clos}\mathcal{P}$ with respect to z_f. Hence continuity of z is a consequence of $z_c(t) = z_f(t)$ for all $t \in \partial\mathcal{P}$. Since each $t \in \partial\mathcal{P}$ satisfies $\sigma(t, z_c(t)) = 0$ it follows $\Psi(t, z_c(t), \cdot) \equiv 0$. Let us assume for a moment that the differential inclusion (DI$_c$) holds for this t. We then find

$$0 \in D\mathcal{E}(t, z_c(t)) + \mathcal{N}_{\partial\mathcal{A}}(z_c(t)) \subset \mathbb{R}^{3^*}. \tag{4.16}$$

For the set of constraint forces we find $\mathcal{N}_{\partial\mathcal{A}}(z_c(t)) = \begin{pmatrix} 0 & 0 & \mathbb{R}^* \end{pmatrix} \subset \mathbb{R}^{3^*}$ and hence (4.16) together with $\sigma(t, z_c(t)) = 0$ imply $D\mathcal{E}(t, z_c(t)) = 0 \in \mathbb{R}^{3^*}$. Hence, $z_c(t)$ satisfies the friction-free differential inclusion (DI$_f$), too. And since $\mathcal{E}(t, \cdot)$ is strictly convex $z_f(t)$ is uniquely determined by (DI$_f$) and we find $z_c(t) = z_f(t)$. Even if (DI$_c$) does not hold for the above $t \in \partial\mathcal{P}$, so we still can establish (4.16) using the continuity of $D\mathcal{E}$ and σ. ∎

Theorem 4.17 (Solution candidate presents a solution) *Let* (O0), (O1) *and* (O4). *For given* $Z_0 \in \partial\mathcal{A}$ *let* $z : [T_0, T_0+\Delta] \to \mathcal{A}$ *be a solution candidate. Then there exists* $\Delta_* \in (0, \Delta]$ *such that* $z \in W^{1,\infty}([T_0, T_0 + \Delta_*], \mathcal{A})$ *and z satisfies, for almost all* $t \in [T_0, T_0 + \Delta_*]$ *the differential inclusion*

$$0 \in D\mathcal{E}(t, z(t)) + \partial_v\Psi(t, z(t), \dot{z}(t)) + \mathcal{N}_{\mathcal{A}}(z(t)) \quad \subset \mathbb{R}^{3^*} \tag{DI}$$

and hence solves Problem 4.1.

Proof: Due to the Regularity Assumption (O1), we can extend the Convexity Assumption (O4) to some neighborhood of (T_0, Z_0) as follows. There exists a time span $\Delta_e \in (0, \Delta]$ and a radius $r > 0$ such that for all $t \in [T_0, T_0 + \Delta_e]$ the function $\mathcal{E}(t, \cdot)$ is uniformly convex on $\mathcal{A} \cap \mathcal{B}_r(Z_0)$, i.e. there exists $\alpha_e > 0$ such that

$$v^\top \mathrm{H}(t, z) v \geq \alpha_e$$

holds for all $(t, z) \in [T_0, T_0 + \Delta_e], z \in \mathcal{B}_r(Z_0) \cap \mathcal{A}$ and all $v \in \mathbb{S}^2$. Due to the Lipschitz continuity of the solutions z_f, z_c there exists a time span $\Delta_* \in (0, \Delta_e]$ such that for all $t \in [T_0, T_0 + \Delta_*]$ we have $z_f(t), z_c(t) \in \mathcal{A} \cap \mathcal{B}_r(Z_0)$ and thus also the solution candidate satisfies $z(t) \in \mathcal{A} \cap \mathcal{B}_r(Z_0)$. By the previous lemma this proves $z \in \mathrm{W}^{1,\infty}([T_0, T_0 + \Delta_*], \mathcal{A})$. The strict convexity of $\mathcal{E}$ will be needed again at the end of the proof, see (4.18).

Our aim is to prove the differential inclusion (DI) for almost all time $t \in [T_0, T_0 + \Delta_*]$. For simplicity we decide to prove the inclusion only for those times t for which the classical derivatives for z and z_c exist, i.e. for all $t \in \mathcal{T}$ with

$$\mathcal{T} := \left\{ t \in [T_0, T_0 + \Delta_*] \; : \; \begin{array}{l} \dot{z}(t) = \lim_{h \to 0} \frac{1}{h}\big(z(t+h) - z(t)\big) \text{ and} \\ \dot{z}_c(t) = \lim_{h \to 0} \frac{1}{h}\big(z_c(t+h) - z_c(t)\big) \end{array} \right\}$$

We define the open set $\mathcal{P} = \{t \in [T_0, T_0 + \Delta_*] \; : \; \sigma(t, z_c(t)) > 0\}$ of times with positive normal force and consider two different cases.

In the first case we assume $t \in \mathcal{T} \cap \mathrm{clos}\mathcal{P}$ and we find $z(t) = z_c(t)$ since both functions are continuous and we have by definition $z \equiv z_c$ on $\mathcal{P}$. We prove now the same for the derivatives. Due to $t \in \mathcal{T}$ both classical derivatives $\dot{z}(t)$ and $\dot{z}_c(t)$ exist. We want to calculate them and choose a sequence $(t_n)_{n \in \mathbb{N}} \subset \mathcal{P}$ with $t_n \to t$ for $n \to \infty$. Further, since the set $\mathcal{P}$ is open, we can assume that $t_n \neq t$ holds for all $n \in \mathbb{N}$. Again by the definition of the solution candidate z we have $z(t_n) = z_c(t_n)$ for all $n \in \mathbb{N}$. Hence, we conclude

$$\dot{z}(t) = \lim_{n \to \infty} \frac{z(t_n) - z(t)}{t_n - t} = \lim_{n \to \infty} \frac{z_c(t_n) - z_c(t)}{t_n - t} = \dot{z}_c(t).$$

This shows that our solution candidate satisfies the differential Inclusion (DI_c) for $t \in \mathcal{T} \cap \mathrm{clos}\mathcal{P}$. Due to the continuity of σ and z_c we find $\sigma(t, z(t)) \geq 0$ for all $t \in \mathrm{clos}\mathcal{P}$. We insert this additional information of having a positive normal force into the Inclusion (DI_c) and replace

$$\mathcal{N} = \begin{pmatrix} 0 & 0 & \mathbb{R}^* \end{pmatrix} \quad \text{by} \quad \mathcal{N}_\mathcal{A}(z(t)) = \begin{pmatrix} 0 & 0 & (-\infty, 0] \end{pmatrix}.$$

This gives us formally the desired inclusion (DI) for $t \in \mathcal{T} \cap \mathrm{clos}\mathcal{P}$.

In the second case we consider $t \in \mathcal{T} \cap (\text{clos}\mathcal{P})^c$. Here $(\text{clos}\mathcal{P})^c$ denotes the complement of the set $\text{clos}\mathcal{P}$. By the definition of our solution candidate z we have $z(t) = z_f(t)$. Hence, the friction-free differential inclusion (DI$_f$) holds for our solution candidate z. To derive the full differential inclusion (DI) we still have to show that in our situation $\partial_v \Psi(t, z(t), \cdot) \equiv 0 \in \mathbb{R}^{3^*}$ holds. This will be a consequence of $\sigma(t, z(t)) = \sigma(t, z_f(t)) \leq 0$. We exclude the opposite and assume

$$\sigma(t, z_f(t)) > 0. \tag{4.17}$$

Due to the definition of the set of normal forces $\mathcal{N}_\mathcal{A}$ this implies $z_f(t) \in \partial\mathcal{A}$. Further, by the strict convexity of the energy functional $\mathcal{E}(t, \cdot)$ on $\mathcal{A} \cap \mathcal{B}_r(Z_0)$, see the beginning of the proof, we find

$$z_f(t) = \text{argmin}\{\mathcal{E}(t, y) : y \in \partial\mathcal{A} \cap \mathcal{B}_r(Z_0)\} \tag{4.18}$$

We compare now $z_f(t)$ and $z_c(t)$. Because of $t \notin \text{clos}\mathcal{P}$ we find $\sigma(t, z_c(t)) \leq 0$ (i.e. $\Psi(t, z_c(t), \cdot) \equiv 0$). Since there is no friction $z_c(t)$ also satisfies (4.18) and we conclude $z_c(t) = z_f(t)$. This implies $\sigma(t, z_f(t)) \leq 0$ which contradicts our Assumption (4.17). Summarizing the second case, we have shown for $t \in \mathcal{T} \cap (\text{clos}\mathcal{P})^c$ that (DI$_f$) and $\partial_v \Psi(t, z(t), \cdot) \equiv 0 \in \mathbb{R}^{3^*}$ hold. Both together imply the desired inclusion (DI). ∎

Corollary 4.18 (Proof of Theorem 4.2 Part 3, using (O5*)) *The above theorem proves the general Existence Theorem 4.2 in the case of having contact and no normal force and under the strong Assumption (O5*)*

Proof: We replace in Theorem 4.2, part 3, the Assumption (O5) by (O5*) and thus assume (O0)–(O4), (O5*), $Z_0 \in \partial\mathcal{A}$ and $\sigma(T_0, Z_0) = 0$. This is sufficient to apply the Existence Theorems 4.4 and 4.11 and to derive the existence of positive time spans $\Delta_f, \Delta_c > 0$ and of friction-free and contact solutions $z_f \in W^{1,\infty}([T_0, T_0 + \Delta_f], \mathcal{A})$ and $z_c \in W^{1,\infty}([T_0, T_0 + \Delta_c], \mathcal{A})$, respectively. We choose $\Delta_{fc} = \min\{\Delta_f, \Delta_c\}$ and define the solution candidate $z \in W^{1,\infty}([T_0, T_0 + \Delta_{fc}], \mathcal{A})$ as in Definition 4.15. Finally, we apply the above theorem and establish the existence of a solution of Problem 4.1. ∎

4.4 Contact subproblem under weak convexity assumption

We prove once again the existence of a solution $z_c \in W^{1,\infty}([T_0, T_0 + \Delta_c], \partial\mathcal{A})$ for the Contact Subproblem 4.8. But this time we replace the strong Convexity Assumption

(O5*) which is based on estimates of norms, i.e.

$$\min_{v \in T_{\partial\mathcal{A}}(Z_0) \cap \mathbb{S}^2} \langle H(T_0, Z_0)v, v \rangle - \sigma(T_0, Z_0)_+ \|DM(Z_0)\| - \|D\sigma(T_0, Z_0)\| \|M(Z_0)\| > 0,$$

(O5*)

by the weaker Joint Convexity Assumption (O5) which is formulated with respect to directions, i.e. for all tangential directions $v \in T_{\partial\mathcal{A}}(Z_0) \cap \mathbb{S}^2$ we assume

$$\langle H(T_0, Z_0)v, v \rangle + \sigma(T_0, Z_0)_+ \frac{(M(Z_0)v)^\top}{\|M(Z_0)v\|} DM(Z_0)[v, v] + \|M(Z_0)v\| \|D\sigma(T_0, Z_0)v > 0.$$

(O5)

We recall the convention $\big(DM(z)[u, v]\big)_i = \sum_{j,k=1}^3 \partial_{z_k} M_{ij}(z) u_j v_k$ for $i = 1, \ldots, 3$ and $u, v \in \mathbb{R}^3$.

Even though in Assumption (O5) it is physically reasonable to consider the change of elastic forces, represented by the term $\langle H(T_0, Z_0)v, v \rangle$, and the change of the frictional forces, represented by the two other terms, with respect to the same sliding direction $v \in T_{\partial\mathcal{A}}(z)$, this causes some technical difficulties for our basic mathematical strategy. There the crucial assumption is $\hat{\alpha}_* > \hat{q}$, which is equivalent to the strong Assumption (O5*) and hence does not hold here. See also the Example 4.20 below for a comparison between (O5) and (O5*). We recall that the strategy is based on a time incremental scheme and the key estimate to derive convergence is the recursive estimate

$$\hat{\alpha}_* \|z_k - z_{k-1}\| \le (\hat{c}_1 + \hat{c}_2) f_\Pi + \hat{q} \|z_{k-1} - z_{k-2}\|$$

which we proved in Lemma 3.18. We see here that two different sliding directions are involved. The recent one, represented by the difference $z_k - z_{k-1}$, and the previous direction $z_{k-1} - z_{k-2}$. This feature of having linked two different directions is typical for our incremental method and we cannot avoid it. Here we face the difficulty of Assumption (O5) considering only a single direction v.

But we will see that for partitions with small fineness f_Π the difference between the two directions $v_1 = \frac{z_k - z_{k-1}}{\|z_k - z_{k-1}\|}$ and $v_2 = \frac{z_{k-1} - z_{k-2}}{\|z_{k-1} - z_{k-2}\|}$ is small in many situations and Assumption (O5) will be sufficient to derive a new but similar recursive estimate. For those situations for which we cannot control, even for small fineness f_Π, the change of directions we locally increase the convexity of $\mathcal{E}(t, \cdot)$ and assure locally (O5*) again. Thus we find the classical recursive estimate again.

The general procedure is as follows: in Section 4.4.1 *increasing the convexity* we introduce parameters $\alpha, \rho > 0$ to increase locally the convexity of the energy. This leads us to a simplified Contact Problem 4.28 with increased convexity. The advantage of the simplified problem is that it can be solved using the weaker Joint Convexity Assumption O5 only, see Section 4.4.2 *solving simplified problems*. For this we modify the scheme of the basic strategy. In fact, we leave the first part *approximative solutions* and the third part *existence* unchanged and only modify the second part *convergence*. In the second part we will derive the two different recursive estimates mentioned above. In the following induction we then use the above classical recursive estimate if we are in a situation with increased convexity and use the new recursive estimate if we are in a situation without increased convexity but small changes of directions. After having established the existence of a solution for the simplified problem we consider in the last Section 4.4.3 *solving the contact problem* a sequence of parameters $(\rho_j)_{j \in \mathbb{N}}$ with $\rho_j \to 0$ and show that in the limit we find our solution.

In the whole section we consider for convenience the equivalent two-dimensional formulation 4.9 of the Contact Problem.

Theorem 4.19 (Existence for contact subproblem) *Assume* (O0)–(O3) *and* (O5) *and* $Z_0 \in \partial\mathcal{A}$. *Then there exists a positive time span* $\Delta_c > 0$ *and a solution* $z_c \in W^{1,\infty}([T_0, T_0 + \Delta_c], \partial\mathcal{A})$ *of the contact subproblem 4.8.*

Proof: The proof is established in Corollary 4.39.   ∎

Before we prove the above theorem we discuss the difference of the two different convexity Assumptions (O5*) and (O5) with an example.

Example 4.20 (Joint convexity vs. convexity with respect to norms)
Assume $A > B > 0$ *to be given. For an arbitrary force* $f \in C^2([0,T], \mathbb{R}^{3^*})$ *we define*

$$\mathcal{E}(t, z) := \frac{1}{2} \left(A z_1^2 (z_3 + 1)^2 + B z_2^2 \right) - f(t)z.$$

We take $T_0 = 0$ *and* $Z_0 = \begin{pmatrix} 1 & 1 & 0 \end{pmatrix}^\top$ *as initial values and calculate for the Hessian matrix*

$$H(T_0, Z_0) = \begin{pmatrix} A & 0 & 2A \\ 0 & B & 0 \\ 2A & 0 & A \end{pmatrix}.$$

Note that the upper left sub-matrix $\bar{H}(T_0, Z_0) \in \mathbb{R}^{2 \times 2}$ *is diagonal. The friction is modeled by a scalar coefficient of friction* $\mu > 0$. *We present in Figure 4.3 the admissible values of* μ *with respect to the variable* $0 < B/A < 1$ *once with respect to the strong Convexity Assumption* (O5*) *and once with respect to the weak Convexity Assumption* (O5). *Considering Figure 4.3 it is clear that* (O5*) *is more restrictive than* (O5). *We now derive the estimates on* μ *with respect to* B/A.
The strong convexity assumption $\alpha_* > q$ *reads*

$$\alpha_* = \min \left\{ v^\top \bar{H}(T_0, Z_0)v \; : \; v \in \mathbb{S}^1 \right\} > \mu \left\| \begin{pmatrix} H_{31}(T_0, Z_0) & H_{32}(T_0, Z_0) \end{pmatrix} \right\| = q.$$

Note that the minimum for the change of elastic forces is attained if the particle slides in the direction $v = \begin{pmatrix} 0 & 1 \end{pmatrix}^\top$ *and we find* $\alpha_* = B$. *The maximal change of*

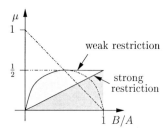

Figure 4.3: Admissible μ for weak and strong Convexity Assumption (O5) and (O5*)

the frictional forces is attained if the particle slides in the perpendicular direction $v = \begin{pmatrix} 1 & 0 \end{pmatrix}^\top$ and we deduce q $= 2\mu A$. *This reveals the clear disadvantage of the strong convexity assumption. Since it based on operator-norms it contains completely different sliding directions while we have physically a single particle which slides in a single direction. All in all the strong Convexity Assumption (O5*) can be expressed equivalently as an assumption on μ, i.e.*

$$\frac{1}{2}\frac{B}{A} > \mu. \tag{4.19}$$

We next discuss the weak Convexity Assumption (O5) with its single sliding direction, i.e.

$$\min\left\{v^\top \bar{\mathsf{H}}(T_0, Z_0)v + \mu \begin{pmatrix} \mathsf{H}_{31}(T_0, Z_0) & \mathsf{H}_{32}(T_0, Z_0) \end{pmatrix} \begin{pmatrix} v_1 & v_2 \end{pmatrix}^\top \ : \ v \in \mathbb{S}^1\right\} > 0. \tag{4.20}$$

In the following we will show that this weak assumption is also equivalent to an estimate on μ, see (4.22) and (4.21) below. In our specific situation (4.20) reads

$$\min\left\{Av_1^2 + Bv_2^2 + 2\mu Av_1 \ : \ v \in \mathbb{S}^1\right\} > 0.$$

Our aim is to express the minimum in dependence on μ. For this we have to know the minimizing direction $v = \begin{pmatrix} v_1 & v_2 \end{pmatrix}^\top \in \mathbb{S}^1$. We substitute $v_2^2 = 1 - v_1^2$ and calculate v_1 via

$$v_1 \in \operatorname{argmin}\left\{(A - B)x^2 + 2\mu Ax + B \ : \ x \in [-1, 1]\right\}.$$

Due to $A > B > 0$ the polynomial is strictly convex and there exists a unique minimizer. We first assume $\mu \geq 1 - B/A$ and find $v_1 = -1$ as minimizer and $A - 2\mu A$ for the minimum. Thus for $\mu \geq 1 - B/A$ we find the restriction

$$\frac{1}{2} > \mu. \tag{4.21}$$

Let us assume $0 < \mu < 1 - B/A$. We consider the derivative of the polynomial and claim $(A - B)v_1 + \mu A = 0$. This leads us to $v_1 = -\mu A/(A - B)$. The above assumption on μ implies $-1 < v_1 < 0$ and hence v_1 is the minimizer. Knowing the minimizer we claim the minimum to be positive, i.e.

$$A\frac{\mu^2 A^2}{(A - B)^2} + B\left(1 - \frac{\mu^2 A^2}{(A - B)^2}\right) - 2\mu A\frac{\mu A}{(A - B)} > 0.$$

After some calculus we find for $\mu < 1 - B/A$ the restriction

$$\sqrt{1 - \frac{B}{A}}\sqrt{\frac{B}{A}} > \mu. \tag{4.22}$$

4.4.1 Increasing the convexity: a simplified problem

In this section we locally increase the convexity of the energy functional. This will lead us to a simplified problem, see Problem 4.28.

We start by introducing new notation since we aim to prove existence of a solution for the two-dimensional Contact Subproblem 4.9 our simplified problem will also formulated with respect to two dimensions. We reformulate the Joint Convexity Assumption (O5) with respect to the two-dimensional setting. In the following we insert a bar ' ‾ ' in the notation whenever the notation refers to the two-dimensional setting and might be confused with the original notation. With the bar ' ‾ ' we want to symbolize the plane $\{z \in \mathbb{R}^3 : z_3 \equiv 0\}$ that is characteristic for the two-dimensional setting. We introduce the normal force $\bar{\sigma}(t, \bar{z}) := \sigma\left(t, \begin{pmatrix} \bar{z}_1 & \bar{z}_2 & 0 \end{pmatrix}^\top\right) \in \mathbb{R}$, the two-dimensional friction matrix $\bar{M}(\bar{z}) := \begin{pmatrix} M_{11} & M_{12} \\ M_{21} & M_{22} \end{pmatrix}\left(\begin{pmatrix} \bar{z}_1 & \bar{z}_2 & 0 \end{pmatrix}^\top\right)$. As in the previous subsection we introduce the energy $\bar{\mathcal{E}} \in C^2([0, T] \times \mathbb{R}^2, \mathbb{R})$ and the dissipation $\bar{\Psi} \in C([0, T] \times \mathbb{R}^2 \times \mathbb{R}^2, [0, \infty))$ via

$$\bar{\mathcal{E}}(t, \bar{z}) := \mathcal{E}\left(t, \begin{pmatrix} \bar{z}_1 & \bar{z}_2 & 0 \end{pmatrix}^\top\right) \quad \text{and}$$

$$\bar{\Psi}(t, \bar{z}, \bar{v}) := \bar{\sigma}(t, \bar{z})_+ \|M(\bar{z})\bar{v}\|.$$

We denote by $\bar{H}(t, \bar{z}) \in \mathbb{R}^{2\times2}$ the Hessian matrix of $\bar{\mathcal{E}}$ with respect to $\bar{z} \in \mathbb{R}^2$. The Joint Convexity Assumption (O5) now reads

$$v^\top\bar{H}(T_0, \bar{z}_0)v + \bar{\sigma}(T_0, \bar{z}_0)\frac{(\bar{M}(\bar{z}_0)v)^\top}{\|\bar{M}(\bar{z}_0)v\|}D\bar{M}(\bar{z}_0)[v, v] + \|\bar{M}(\bar{z}_0)v\|D\bar{\sigma}(T_0, \bar{z}_0)v > 0 \quad (\overline{O5})$$

for all $v \in \mathbb{S}^1$.

After having introduced the two-dimensional notation we want to exploit the continuity of the data $\mathcal{E}$ and M to extend the Assumption ($\overline{O5}$) to some neighborhood

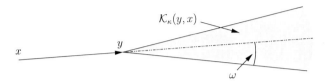

Figure 4.4: Cone $\mathcal{K}_\kappa(y, x)$ with angle ω satisfying $\cos(\omega) = 1 - \kappa$.

of the initial values. We thus provide a playing field on which we construct our solution, see (4.28) and (4.29) for the resulting time span Δ_e and admissible set $\mathcal{A}_e$. But this time we will not only have to extend time T_0 and state $\bar{z}_0$ to some interval $[T_0, T_0+\Delta]$ or ball $\mathcal{B}_r(\bar{z}_0) \subset \mathbb{R}^2$ but we also have to allow for different directions. For this we introduce now the notion of a cone.

Definition 4.21 (Cone) *For given states $x, y \in \mathbb{R}^2$ and given $\kappa \in [0, 1]$ we define the cone $\mathcal{K}_\kappa(y, x) \subset \mathbb{R}^2$ via*

$$\mathcal{K}_\kappa(y, x) := \left\{ z \in \mathbb{R}^2 \ : \ \langle z - y, y - x \rangle \geq \|z - y\|\|y - x\|(1 - \kappa) \right\}.$$

For $0 < \kappa \ll 1$ and for the elements $z \in \mathcal{K}_\kappa(y, x)$ the vector $z - y$ points almost in the direction $y - x$. In fact, let ω denote the angle between the vectors $z - y$ and $y - x$ then we have $\cos(\omega) \geq 1 - \kappa$, see also Figure 4.4. The extension of $(\overline{O5})$ is technical and we use several auxiliary functions to express it. We define the functions $a \in C([0, T] \times \mathbb{R}^2 \times \mathbb{S}^1, \mathbb{R})$ and $q \in C([0, T] \times \mathbb{R}^6 \times \mathbb{S}^1 \times \mathbb{S}^1, \mathbb{R})$ via

$$a(t_1, z_1, v) := \langle v, \bar{H}(t_1, z_1)v \rangle \quad \text{and}$$

$$q(t_2, z_2, z_3, z_4, v, w) := D\bar{\sigma}(t_2, z_2)w\|\bar{M}(z_3)v\| + \bar{\sigma}(t_2, z_4)_+ \frac{(\bar{M}(z_2)v)^\top}{\|\bar{M}(z_2)v\|} D\bar{M}(z_2)\,[v, w].$$

The Assumption $(\overline{O5})$ can now be reformulated as

$$\inf \left\{ a(T_0, \bar{z}_0, v) + q(T_0, \bar{z}_0, \bar{z}_0, \bar{z}_0, v, v) \ : \ v \in \mathbb{S}^1 \right\} > 0.$$

Since we have $a(t, z, v) = a(t, z, -v)$ and $q(t, z, z, z, v, v) = -q(t, z, z, z, -v, -v)$ we see that $(\overline{O5})$ is also equivalent to

$$\inf \left\{ a(T_0, \bar{z}_0, v) - |q(T_0, \bar{z}_0, \bar{z}_0, \bar{z}_0, v, v)| \ : \ v \in \mathbb{S}^1 \right\} > 0.$$

Let us now assume (O5) or equivalently $(\overline{O5})$. We start to extend this assumption to some neighborhood of the arguments. We first introduce for $\Delta \in [0, T - T_0], r \in [0, \infty)$ and $v, w \in \mathbb{S}^1$ the functions

$$\underline{a}(\Delta, r, v) := \inf \left\{ a(t_1, z_1, v) \ : \ t \in [T_0, T_0+\Delta], z_1 \in \mathcal{B}_r(\bar{z}_0) \right\} \quad \text{and} \qquad (4.23)$$

$$\overline{q}(\Delta, r, v, w) := \sup \left\{ q(t, z_2, z_3, z_4, v, w) \ : \ t \in [T_0, T_0+\Delta], z_{2,3,4} \in \mathcal{B}_r(\bar{z}_0) \right\} \qquad (4.24)$$

Note that $(\overline{O5})$ is equivalent to $\inf\{\underline{a}(0,0,v) - \bar{q}(0,0,v,v) : v \in \mathbb{S}^1\} > 0$. We next introduce for $\kappa \in [0,1]$ the function

$$g(\Delta, r, w, \kappa) := \inf\{\underline{a}(\Delta, r, v) - \underline{q}(\Delta, r, v, w) : v \in \mathcal{K}_\kappa(0, -w) \cap \mathbb{S}^1\}. \quad (4.25)$$

Due to the regularity Assumptions (O1) and (O2) on $\mathcal{E}$ and M the function $\underline{a}, \underline{q}$ and g are continuous. Minimizing g with respect to w leads us to the function $G \in C([0, T - T_0] \times [0, \infty) \times [0, 1], \mathbb{R})$ via

$$G(\Delta, r, \kappa) := \inf\{g(\Delta, r, w, \kappa) : w \in \mathbb{S}^1\} \quad (4.26)$$

We conclude that Assumption $(\overline{O5})$ is equivalent to $G(0,0,0) > 0$. Further the function G is in every variable monotone decreasing and since it is continuous there exists an extension triple $(\Delta, r, \kappa) \in (0, \infty)^3$ such that

$$G(\Delta, r, \kappa) > 0 \quad \text{holds.}$$

We will call this inequality an extended version of the joint convexity assumption. We summarize this result.

Lemma 4.22 (Existence of an extension of (O5)**)** *Let* (O1), (O2) *and* (O5). *Then there exists an extension triple* $(\Delta, r, \kappa) \in (0, T-T_0] \times (0, \infty) \times (0, 1]$ *such that*

$$G(\Delta, r, \kappa) > 0. \quad (4.27)$$

Further the function $\bar{\mathcal{E}}(t, \cdot)$ *is uniformly convex on* $\mathcal{B}_r(\bar{z}_0)$ *for all times* $t \in [T_0, T_0 + \Delta]$.

Proof: The extended joint convexity (4.27) implies

$$\bar{\alpha} := \inf\{\langle v, \bar{H}(t, z)v\rangle : t \in [T_0, T_0 + \Delta], z \in \mathcal{B}_r(\bar{z}_0), v \in \mathbb{S}^1\} > G(\Delta, r, \kappa) > 0.$$

This proves the $\bar{\alpha}$-uniform convexity of $\bar{\mathcal{E}}(t, \cdot)$. See also Lemma 3.5. ∎

In the remaining part of this section and in the following sections we will always assume (O1), (O2) and (O5) and we fix an extension triple (Δ, r, κ) satisfying (4.27). We introduce a notation with index 'e' for extension. To make clear that we assume this triple to remain unchanged for the rest of this chapter we write

$$\Delta_e := \Delta > 0 \quad (4.28)$$

to denote the time span of the fixed extension triple. Further we choose a radius $r_e \in (0, r]$ such that $\min\{\bar{M}(\bar{z}) : \bar{z} \in \mathcal{B}_{r_e}(\bar{z}_0), v \in \mathbb{S}^1\} > 0$ holds. This is an extension of the Assumption $\min\{\|M(Z_0)v\| : v \in \mathcal{T}_{\partial\mathcal{A}}(Z_0) \cap \mathbb{S}^2\} > 0$. We define the admissible set of the extension via

$$\mathcal{A}_e := \mathcal{B}_{r_e}(\bar{z}_0) \quad \subset \mathbb{R}^2. \quad (4.29)$$

We introduce the constant $g_e := G(\Delta_e, r_e, \kappa)$ and call the inequality

$$g_e > 0 \qquad (4.30)$$

the extended joint convexity. As a direct consequence we find $\bar{\alpha}_e \geq g_e > 0$ for the constant

$$\bar{\alpha}_e := \min\left\{\underline{a}(\Delta_e, r_e, v) \ : \ v \in \mathbb{S}^1\right\} \qquad (4.31)$$
$$= \min\left\{v^\top \bar{H}(t, z)v \ : \ t \in [T_0, T_0 + \Delta_e], z \in \mathcal{B}_{r_e}(\bar{z}_0), v \in \mathbb{S}^1\right\}.$$

With the choice of $\mathcal{A}_e$ and $[T_0, T_0 + \Delta_e]$ we have now determined our local neighborhood of the initial values $\bar{z}_0$ and T_0 for which we want to construct a solution of the two-dimensional Contact Subproblem 4.9. We thus have determined our playing field for the rest of the chapter.

Next, we want to increase the convexity of the energy functional $\bar{\mathcal{E}}$ in a neighborhood of the pairs $(t, z) \in [T_0, T_0 + \Delta_e] \times \mathcal{A}_e$ satisfying $D\bar{\mathcal{E}}(t, z) = 0 \in \mathbb{R}^{2^*}$. These pairs will be included in the graph of the function $\bar{z}_f$ defined now.

Lemma 4.23 (Unique energy minimizer in $\mathcal{A}_e$) *If* (O1), (O2) *and* (O5) *hold. Then the minimizing problem*

$$\bar{z}_f(t) = \operatorname{argmin}\left\{\bar{\mathcal{E}}(t, y) \ : \ y \in \mathcal{A}_e\right\} \quad \text{for all } t \in [T_0, T_0 + \Delta_e] \qquad (4.32)$$

defines a unique function $\bar{z}_f \in W^{1,\infty}\left([T_0, T_0 + \Delta_e], \mathcal{A}_e\right)$.

Proof: The existence and uniqueness of the function $\bar{z}_f$ follows directly from the $\bar{\alpha}_e$-uniform convexity of $\bar{\mathcal{E}}$, see Lemma 4.22 and (4.31). To prove $\bar{z}_f \in W^{1,\infty}([T_0, T_0 + \Delta_e], \mathcal{A}_e)$ we construct the function again as a solution of Problem 3.1 with the choice $\hat{\mathcal{E}} = \bar{\mathcal{E}}$ and $\hat{\Psi} \equiv 0$ there. Equation (4.32) is then equivalent to the stability (S) in Problem 3.1 and the existence and Lipschitz continuity of a solution follows from Theorem 3.2 ∎

As mentioned above $D\bar{\mathcal{E}}(t, z) = 0$ implies $z = \bar{z}_f(t)$ for $t \in [T_0, T_0 + \Delta_e]$ and $z \in \mathcal{A}_e$. This is important since exactly the pairs (t, z) satisfying $D\bar{\mathcal{E}}(t, z) = 0$ cause technical difficulties if we work only with the weak Joint Convexity Assumption (O5). But we will see later that we can avoid these difficulties if we keep an arbitrary small distance to these pairs. This motivates the introduction of a distance parameter $\rho > 0$ and of the notion of a tube.

Definition 4.24 (Tube) *For a given parameter $\rho > 0$ and for the energy minimizer $\bar{z}_f \in W^{1,\infty}([T_0, T_0 + \Delta_e], \mathcal{A}_e)$ defined by (4.32) we define the tube*

$$T_\rho := \{(t, z) \in [T_0, T_0 + \Delta_e] \times \mathcal{A}_e \ : \ \|z - \bar{z}_f(t)\| \leq \rho\}.$$

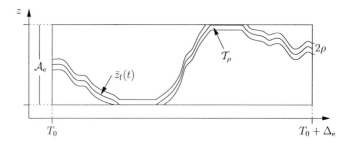

Figure 4.5: Tube $\mathcal{T}_\rho$ of Definition 4.24.

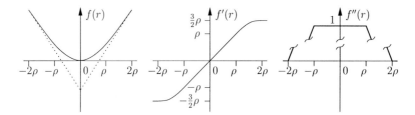

Figure 4.6: The auxiliary function f.

In later sections the parameter ρ will tend to zero and hence the tube shrinks to the graph of $\bar{z}_f$ in the limit. As mentioned above we will be able to construct those parts of the solution remaining outside of a tube using the weak Assumption (O5) only. But for all parts of the solution lying inside of the tube we need the strong Assumption (O5*). Our aim is to introduce a modified energy functional, see $\bar{\mathcal{E}}_\rho$ below, with an increased convexity inside of the Tube $\mathcal{T}_\rho$ such that the strong Assumption (O5*) holds inside of the tube. For this we define a convex auxiliary function $f_\rho \in C^2(\mathbb{R}, \mathbb{R})$ via

$$f_\rho(r) := \begin{cases} \frac{1}{2}r^2 & \text{for } |r| \leq \rho, \\ -\frac{1}{6\rho}|r|^3 + r^2 - \frac{\rho}{2}|r| + \frac{\rho^2}{6} & \text{for } \rho \leq |r| \leq 2\rho \quad \text{and} \\ \frac{3}{2}\rho|r| - \frac{7}{6}\rho^2 & \text{for } 2\rho \leq |r|. \end{cases} \quad (4.33)$$

The construction of this function f_ρ is such that its second derivative f_ρ'' is continuous and satisfies $f_\rho''(z) \in [0,1]$ with $f_\rho''(z) = 1$ for $|z| \leq \rho$. See also Figure 4.6. We use this function to increase the convexity of $\bar{\mathcal{E}}$ within the tube $\mathcal{T}_\rho$.

Definition 4.25 (Energy with locally increased convexity) *For given $\alpha, \rho > 0$ we define the energy functional $\bar{\mathcal{E}}_\rho \in C^2([T_0, T_0 + \Delta_e] \times \mathcal{A}_e, \mathbb{R})$ for all $t \in [T_0, T_0 +$*

$\Delta_e]$ and $z \in \mathcal{A}_e$ by

$$\bar{\mathcal{E}}_\rho(t,z) := \bar{\mathcal{E}}(t,z) + \alpha \left(\sum_{j=1}^2 f_\rho\big((z - \bar{z}_f(t))_j\big) - \sum_{j=1}^2 f_\rho{}'\big((\bar{z}_0 - \bar{z}_f(T_0))_j\big) z_j \right)$$

and with the function $f_\rho \in C^2(\mathbb{R}, \mathbb{R})$ as defined in (4.33).

The second sum in the additional term is such that the initial force remains unchanged. We have

$$D\bar{\mathcal{E}}_\rho(T_0, \bar{z}_0) = D\bar{\mathcal{E}}(T_0, \bar{z}_0) \quad \in \mathbb{R}^{2^*}. \tag{4.34}$$

With the new parameter $\alpha > 0$ we will control the additional convexity, see also Lemma 4.26 below. We specify the choice of α in (4.40). Hence, in contrast to $\rho > 0$ the parameter α remains unchanged throughout this chapter. That is why we represent only ρ in the notation $\bar{\mathcal{E}}_\rho$.

Lemma 4.26 (Increased α-uniform convexity of $\bar{\mathcal{E}}_\rho$ inside of tube) *Let the Assumptions (O1), (O2) and (O5) hold and assume $\alpha, \rho > 0$ to be given. If we denote by $\bar{H}_\rho(t,z) \in \mathbb{R}^{2 \times 2}$ the Hessian matrix of $\bar{\mathcal{E}}_\rho$ then we find*

$$\min \left\{ v^\top \bar{H}_\rho(t,z) v \; : \; (t,z) \in \mathcal{T}_\rho, v \in \mathbb{S}^1 \right\} > \alpha. \tag{4.35}$$

Proof: Due to the Definition 4.25 of $\bar{\mathcal{E}}_\rho$ we have

$$\bar{H}_\rho(t,z) = \bar{H}(t,z) + \alpha \begin{pmatrix} f_\rho{}''\big((z - \bar{z}_f(t))_1\big) & 0 \\ 0 & f_\rho{}''\big((z - \bar{z}_f(t))_2\big) \end{pmatrix}$$

and since $(t,z) \in \mathcal{T}_\rho$ implies $|(z - \bar{z}_f(t))_j| \leq \rho$ for $j = 1, 2$ we find $f_\rho{}''\big((z - \bar{z}_f(t))_1\big) = 1$. Together with the uniform convexity of $\bar{\mathcal{E}}(t, \cdot)$ on the whole set $\mathcal{A}_e$ and for all times $t \in [T_0, T_0 + \Delta_e]$, see Lemma 4.22, this establishes the strict inequality. ∎

We have now introduced our admissible set $\mathcal{A}_e \subset \mathbb{R}^2$, time interval $[T_0, T_0 + \Delta_e]$, the dissipation functional $\bar{\Psi}$ and determined the new energy $\bar{\mathcal{E}}_\rho$ up to the constant α. We next verify as many Assumptions (3.1)–(3.11) of the basic mathematical strategy as possible and fix at the end α in (4.40).

The Assumptions (3.1)–(3.6) follow directly from (O0)–(O2) and (O5). The Assumption (3.11) on the initial values $T_0, \bar{z}_0$ is due to the convexity of $\bar{\mathcal{E}}_\rho$ equivalent to

$$0 \in D\bar{\mathcal{E}}_\rho(T_0, \bar{z}_0) + \partial_v \bar{\Psi}(T_0, \bar{z}_0, 0) + \mathcal{N}_{\mathcal{A}_e}(\bar{z}_0) \quad \subset \mathbb{R}^{2^*},$$

see Lemma 2.8 for the equivalence. Since $\mathcal{A}_e = \mathcal{B}_{r_e}(\bar{z}_0)$ holds we find $\mathcal{N}_{\mathcal{A}_e}(\bar{z}_0) = 0 \in \mathbb{R}^{2^*}$. Further we have chosen $\bar{\mathcal{E}}_\rho$ such that $D\bar{\mathcal{E}}_\rho(T_0, \bar{z}_0) = D\bar{\mathcal{E}}(T_0, \bar{z}_0)$ and hence the above inclusion is equivalent to

$$0 \in D\bar{\mathcal{E}}(T_0, \bar{z}_0) + \partial_v \bar{\Psi}(T_0, \bar{z}_0, 0) \quad \subset \mathbb{R}^{2^*}.$$

But these are exactly the first two lines of the Assumption (O3). Hence, up to now our data $\mathcal{A}_e, \bar{\mathcal{E}}_\rho$ and $\bar{\Psi}$ satisfies the Assumptions (3.1)–(3.6) and (3.11) and we still miss the estimates (3.7)–(3.10). For this we define the cylinder

$$\mathcal{C}_e := [T_0, T_0 + \Delta_e] \times \mathcal{A}_e \qquad (4.36)$$

and introduce the constants

$$\bar{c}_1 := \frac{1}{2} \left(\left\| \partial_t D\bar{\mathcal{E}} \right\|_{\mathrm{L}^\infty(\mathcal{C}_e, \mathbb{R}^{2*})} + \alpha \left\| \dot{\bar{z}}_f \right\|_{\mathrm{L}^\infty([T_0, T_0 + \Delta_e], \mathbb{R}^2)} \right), \qquad (4.37)$$

$$\bar{c}_2 := \left\| \bar{M} \right\|_{\mathrm{L}^\infty(\mathcal{C}_e, \mathbb{R}^{2 \times 2})} \left\| \partial_t \bar{\sigma} \right\|_{\mathrm{L}^\infty(\mathcal{C}_e, \mathbb{R})} \quad \text{and} \qquad (4.38)$$

$$\bar{q} := \left\| D\bar{\sigma} \right\|_{\mathrm{L}^\infty(\mathcal{C}_e, \mathbb{R}^{2*})} \left\| \bar{M} \right\|_{\mathrm{L}^\infty(\mathcal{A}_e, \mathbb{R}^{2 \times 2})} + \left\| \bar{\sigma} \right\|_{\mathrm{L}^\infty(\mathcal{C}_e, \mathbb{R})} \left\| D\bar{M} \right\|_{\mathrm{L}^\infty(\mathcal{A}_e, \mathbb{R}^{2 \times 2 \times 2})}. \qquad (4.39)$$

The constants $\bar{c}_1, \bar{c}_2$ and $\bar{q}$ correspond to those found in the Assumptions (3.7)–(3.9). Note that we use $f_\rho'' \in [0, 1]$ for the estimate of $\bar{c}_1$. We summarize.

Lemma 4.27 *Let $\alpha, \rho > 0$ be given. The Assumptions (O0)–(O3) and (O5) on the set $\mathcal{A} \subset \mathbb{R}^3$ and the functionals $\mathcal{E}$ and Ψ imply that the two-dimensional data $\mathcal{A}_e \subset \mathbb{R}^2$, $\bar{\mathcal{E}}_\rho$ and $\bar{\Psi}$ satisfies the Assumptions (3.1)–(3.9) and (3.11). The uniform convexity (3.3) of $\bar{\mathcal{E}}_\rho$ holds for the constant $\bar{\alpha}_e > 0$ as defined in (4.31).*

Note that $\bar{\alpha}_e$ in general does not satisfy $\bar{\alpha}_e > \bar{q}$. But since the constant $\bar{q}$ is independent of the parameter α we choose $\alpha > 0$ big enough such that

$$\alpha > \bar{q}. \qquad (4.40)$$

Hence by this choice of the parameter α we increase the uniform convexity of $\bar{\mathcal{E}}_\rho$ inside of the tube $\mathcal{T}_\rho$ such that for $(t, z) \in \mathcal{T}_\rho, v \in \mathbb{S}^1$ we find

$$v^\top \bar{H}_\rho(t, z) v \geq \alpha > \bar{q}.$$

This is formally (O5*). Thus we finally achieved our aim and presented modified data $\bar{\mathcal{E}}_\rho$ and $\bar{\Psi}$ such that the strong Assumption (O5*) (or equivalently (3.10)) holds locally within some region, i.e the tube. We now present the corresponding problem to solve.

Problem 4.28 (Contact subproblem with increased convexity)
For α as in (4.40) and for given $\rho > 0$ find a time span $\Delta \in (0, \Delta_e]$, which is independent of ρ, and a function $z_\rho \in \mathrm{W}^{1,\infty}([T_0, T_0 + \Delta], \mathcal{A}_e)$ such that the initial condition $z_\rho(T_0) = \bar{z}_0$ is satisfied and such that for almost all $t \in [T_0, T_0 + \Delta]$ we have

$$0 \in D\bar{\mathcal{E}}_\rho(t, z_\rho(t)) + \partial_v \bar{\Psi}(t, z_\rho(t), \dot{z}_\rho(t)) + \mathcal{N}_{\mathcal{A}_e}(z_\rho(t)) \quad \subset \mathbb{R}^{2*}. \qquad (\mathrm{DI}_\rho)$$

4.4.2   Solving simplified problems

In this section we assume the parameter $\rho > 0$ to be given and construct a solution z_ρ for a single simplified Problem 4.28.

The construction follows mainly the basic mathematical strategy, which consists of three parts. The parts *approximative solutions* and *existence* are left unchanged. New ideas are needed for the *convergence* part since in the basic strategy we used in this part the strong Convexity Assumption 3.10 or equivalently (O5*). We constructed the simplified problem such that this strong assumption holds at least inside of the tube $\mathcal{T}_\rho$. We show in this section that outside of a tube the weak Assumption (O5) is sufficient.

We want to recall that in the previous section we determined our playing field $[T_0, T_0 + \Delta_e] \times \mathcal{A}_e$ in (4.28) and (4.29) and we fixed the parameter $\alpha > 0$ in (4.40). These constants remain unchanged throughout the whole chapter and we will not recall their definitions in the following.

Definition 4.29 (Incremental Problem and approximative solutions) *For given initial value $z_0 = \bar{z}_0 \in \mathcal{A}_e$, time span $\Delta \in (0, \Delta_e]$ and an equi-distant partition $\Pi : T_0 = t_0 < t_1 < \cdots < t_{N_\Pi} = T_0 + \Delta$ the Incremental Problem consists in finding incrementally $z_1, z_2, z_3, \ldots, z_{N_\Pi}$ such that for $k = 1, \ldots, N_\Pi$*

$$z_k \in \operatorname{argmin} \left\{ \bar{\mathcal{E}}_\rho(t_k, y) + \bar{\Psi}(t_{k-1}, z_{k-1}, y - z_{k-1}) \; : \; y \in \mathcal{A}_e \right\}. \qquad (\mathrm{IP}_\rho)$$

For a sequence of equi-distant partitions $(\Pi^{(n)})_{n \in \mathbb{N}}$ of $[T_0, T_0 + \Delta]$, i.e.

$$\Pi^{(n)} : T_0 = t_0^{(n)} < t_1^{(n)} < \cdots < t_{N_{\Pi^{(n)}}}^{(n)} = T_0 + \Delta$$

we define the sequence of approximative solutions $(z_\rho^{(n)})_{n \in \mathbb{N}} \subset \mathrm{W}^{1,\infty}([T_0, T_0 + \Delta], \mathcal{A}_e)$ as the piecewise linear interpolation between the values $(z_k^{(n)})_{k=1,\ldots,N_{\Pi^{(n)}}}$ of the solution of the corresponding Incremental Problem (IP_ρ), i.e.

$$z_\rho^{(n)}(t) := z_{k-1}^{(n)} + \frac{t - t_{k-1}^{(n)}}{t_k^{(n)} - t_{k-1}^{(n)}} (z_k^{(n)} - z_{k-1}^{(n)}) \quad \text{for } t \in [t_{k-1}^{(n)}, t_k^{(n)}] \text{ and } k = 1, \ldots, N_{\Pi^{(n)}}.$$

Theorem 4.30 (Existence and uniqueness of approximative solutions) *Let the Assumptions (O0)–(O3) and (O5) hold. Then for any $\Delta \in (0, \Delta_e]$ and any arbitrary sequence of partitions $(\Pi^{(n)})_{n \in \mathbb{N}}$ of $[T_0, T_0 + \Delta]$ the corresponding sequence of approximative solutions $(z_\rho^{(n)})_{n \in \mathbb{N}} \subset \mathrm{W}^{1,\infty}([T_0, T_0 + \Delta], \mathcal{A}_e)$ of (IP_ρ) exists and is unique.*

Proof: The existence and uniqueness of approximative solutions was already proven in Lemma 3.12. In Lemma 4.27 we verified all necessary assumptions on our tow-dimensional data $\mathcal{A}_e$, $\bar{\mathcal{E}}_\rho$ and $\bar{\Psi}$.   ∎

This completes the part *approximative solutions* of the basic strategy. The following *convergence* part has to be modified.

The aim of this part is to establish convergence of the approximative solutions by showing that they are uniform Lipschitz continuous. The key argument for uniform Lipschitz continuity is a recursive estimate of the form

$$\|z_k - z_{k-1}\| \leq C|t_k - t_{k-1}| + R\|z_{k-1} - z_{k-2}\|$$

for the discrete solution $(z_k)_{k=0,\ldots,N_\Pi}$ and constants $C, R \geq 0$ being independent of Π. Depending on (t_{k-1}, z_{k-1}) being inside or outside to the tube $\mathcal{T}_\rho$ we derive two different estimates of this type, see (4.41) and (4.42) below. The advantage of this distinction is that outside of the tube the directions of the vectors $z_k - z_{k-1}$ and $z_{k-1} - z_{k-2}$ change only little, see Proposition 4.35. This allows us to establish outside of the tube the recursive estimate (4.42) using the direction sensitive weak Assumption (O5) only.

We do not have to reformulate the whole convergence part of the basic strategy. Our starting point is Lemma 3.17 with its estimate on minimizing problems. We will replace Lemma 3.18 and its recursive estimate by the two following Propositions 4.31 and 4.33.

Proposition 4.31 (Lipschitz estimate inside of tube) *Let us assume* (O0)–(O3) *and* (O5). *Further, we assume the times* $T_0 \leq \tau_0 \leq \tau_1 \leq \tau_2 \leq T_0 + \Delta_e$ *and the value* $z_0 \in \mathcal{A}_e$ *to be given. We define first* $z_1 \in \mathcal{A}_e$ *and then* $z_2 \in \mathcal{A}_e$ *by*

$$z_1 \in \operatorname{argmin}\left\{\bar{\mathcal{E}}_\rho(\tau_1, y) + \bar{\Psi}(\tau_0, z_0, y - z_0) : y \in \mathcal{A}_e\right\} \quad \text{and}$$
$$z_2 \in \operatorname{argmin}\left\{\bar{\mathcal{E}}_\rho(\tau_2, y) + \bar{\Psi}(\tau_1, z_1, y - z_1) : y \in \mathcal{A}_e \cap \mathcal{B}_{\rho/4}(z_1)\right\}.$$

If for the cylinder $\mathcal{C} := [\tau_0, \tau_2] \times \mathcal{A}_e \cap \mathcal{B}_{\rho/4}(z_1)$ *we find* $\mathcal{C} \subset \mathcal{T}_\rho$ *and* $(\tau_0, z_0) \in \mathcal{C}$, *then the recursive estimate*

$$\alpha\|z_2 - z_1\| \leq (\bar{c}_1 + \bar{c}_2)\max\{|\tau_2 - \tau_1|, |\tau_1 - \tau_0|\} + \bar{q}\|z_1 - z_0\| \tag{4.41}$$

holds with the constants $\bar{c}_1, \bar{c}_2, \bar{q}$ *and* α *as defined in* (4.37)–(4.39) *and* (4.40).

Remark 4.32 *Due to our choice* $\alpha > \bar{q}$ *in* (4.40) *we find*

$$\|z_2 - z_1\| \leq C\max\{|\tau_2 - \tau_1|, |\tau_1 - \tau_0|\} + R\|z_1 - z_0\|$$

with $0 \leq R < 1$.

Proof: The recursive estimate (4.41) is of the same type as the third estimate (3.25) for minimizing problems which we proved in Lemma 3.18. Hence, we have to check the Assumption (3.1)–(3.9) of 3.18 for our current data. Since we are inside of the tube $\mathcal{T}_\rho$ we find $\bar{\mathcal{E}}_\rho$ to be α-uniformly convex with the constant α as defined in (4.40). To clarify that we are in the same situation we adapt some of the notation of Lemma 3.18. We put $\hat{T} = \tau_0$ and $\hat{\Delta} = \tau_2 - \tau_0$ such that $[\hat{T}, \hat{T} + \hat{\Delta}] = [\tau_0, \tau_2]$ and further define $\hat{\mathcal{A}} := \mathcal{A}_e \cap \mathcal{B}_{\rho/4}(z_1)$. We assume $(\tau_0, z_0) \in \mathcal{C}$ and thus $z_0 \in \hat{\mathcal{A}}$. Further we have by definition $z_2 \in \hat{\mathcal{A}}$. Obviously we can replace the set $\mathcal{A}_e$ by $\hat{\mathcal{A}}$ in the definition of z_1 and thus are formally in the situation of Lemma 3.18.

We now check the assumptions. The set $\hat{\mathcal{A}} \subset \mathbb{R}^2$ is closed and convex. This is the Assumption (3.1). Our assumption $\mathcal{C} \subset \mathcal{T}_\rho$ together with the α-uniform convexity of $\bar{\mathcal{E}}_\rho$ inside of the tube $\mathcal{T}_\rho$, which we have shown in Lemma (4.26), imply

$$\min\left\{ v^\top \bar{H}_\rho(\tau, z)v \ : \ v \in \mathbb{S}^1 \right\} \geq \alpha \quad \text{for all } (\tau, z) \in [\hat{T}, \hat{T} + \hat{\Delta}] \times \hat{\mathcal{A}}.$$

This is exactly the Assumption (3.3) that we need for Lemma 3.18. The rest of the assumptions we already verified in Lemma 4.27. Hence, applying Lemma 3.18 establishes our result. ∎

Especially the Proposition 4.33 is an improvement of Lemma 3.18 or Theorem 3.19 since it proves Lipschitz continuity without using the strong Assumption (3.10) or equivalently (O5*).

Proposition 4.33 (Lipschitz estimate inside of a cone) *Let the Assumptions* (O0)–(O3) *and* (O5) *hold. Further assume the times* $\tau_1, \tau_2, t_1, t_2 \in [T_0, T_0 + \Delta_e]$ *and the value* $z_0 \in \mathcal{A}_e$ *to be given and define first* $z_1 \in \mathcal{A}_e$, *then choose any closed convex set* $\mathcal{M}(z_1) \subset \mathbb{R}^2$ *with* $z_1 \in \mathcal{M}$ *and finally define* $z_2 \in \mathcal{A}_e$ *as follows*

$$z_1 \in \operatorname{argmin}\left\{ \bar{\mathcal{E}}_\rho(\tau_1, y) + \bar{\Psi}(t_1, z_0, y - z_0) \ : \ y \in \mathcal{A}_e \right\} \quad \text{and}$$
$$z_2 \in \operatorname{argmin}\left\{ \bar{\mathcal{E}}_\rho(\tau_2, y) + \bar{\Psi}(t_2, z_1, y - z_1) \ : \ y \in \mathcal{A}_e \cap \mathcal{M}(z_1) \cap \mathcal{K}_\kappa(z_1, z_0) \right\}.$$

Further we assume $\|z_2 - z_1\| > \|z_1 - z_0\|$. *Then the following estimate holds*

$$\|z_2 - z_1\| \leq \frac{\bar{c}_1 + \bar{c}_2}{g_e} \max\left\{ |\tau_2 - \tau_1|, |t_2 - t_1| \right\} \tag{4.42}$$

with the constants $\bar{c}_1, \bar{c}_2$ *and* g_e *as defined in* (4.37), (4.38) *and* (4.30).

Remark 4.34 *For the time being the set* $\mathcal{M}$ *is not needed, i.e. we may choose* $\mathcal{M}(z_1) = \mathbb{R}^2$. *But later on we will use nontrivial* $\mathcal{M}$, *see* (4.45).

Proof: The existence and uniqueness of the minimizers $z_1, z_2 \in \mathcal{A}_e$ both follow from Proposition 3.10. Since $z_1 \in \mathcal{K}_\kappa(z_1, z_0)$ holds we see that z_1 is a minimizer with respect to the set $\hat{\mathcal{A}} = \mathcal{A}_e \cap \mathcal{M}(z_1) \cap \mathcal{K}_\kappa(z_1, z_0)$, too.

Our starting point is Lemma 3.17 with exactly this choice for the set $\hat{\mathcal{A}}$. The set $\hat{\mathcal{A}}$ is closed and convex. We verified the rest of the assumptions of Lemma 3.17 in Lemma 4.27. But before recalling the estimate (3.22) of this lemma we want to emphasize that we are interested in directions as they appear in the Assumption (O5). Hence we make use of our assumption $\|z_2 - z_1\| > \|z_1 - z_0\|$, divide the inequality (3.22) by $\|z_2 - z_1\| > 0$, and introduce the direction $v = \frac{z_2 - z_1}{\|z_2 - z_1\|}$. We then find

$$\int_0^1 \int_0^r \langle v, \left(\bar{H}_\rho(\tau_1, y(s)) + \bar{H}_\rho(\tau_2, y(1-s)) \right) v \rangle \, ds \, dr \, \|z_2 - z_1\|$$
$$\leq \int_0^1 \int_0^r \partial_t D \bar{\mathcal{E}}_\rho(\tau(r), y(s)) v \, ds \, dr (\tau_2 - \tau_1) + \bar{\Psi}(t_1, z_0, v) - \bar{\Psi}(t_2, z_1, v)$$

for the definitions $y(s) := z_1 + s(z_2 - z_1)$ and $\tau(r) := \tau_1 + r(\tau_2 - \tau_1)$.

The double integral on the left hand side is estimated from below by $\underline{a}(\Delta_e, r_e, v)$, see also the definition of $\underline{a}$ in (4.23). In fact the definition of $\underline{a}$ was exactly chosen such that this estimate holds. The double integral on the right hand side is estimated from above by $\bar{c}_1 |\tau_2 - \tau_1|$ and with the constant $\bar{c}_1$ defined in (4.37). We expand the difference of dissipations as follows $\bar{\Psi}(t_1, z_0, v) - \bar{\Psi}(t_2, z_0, v) + \bar{\Psi}(t_2, z_0, v) - \bar{\Psi}(t_2, z_1, v)$. The first of these two differences is estimated by $\bar{c}_2 |t_2 - t_1|$ while the second difference will need a far more careful consideration. We summarize

$$\underline{a}(\Delta_e, r_e, v) \|z_2 - z_1\|$$
$$\leq (\bar{c}_1 + \bar{c}_2) \max \left\{ |\tau_2 - \tau_1|, |t_2 - t_1| \right\} + \bar{\Psi}(t_2, z_0, v) - \bar{\Psi}(t_2, z_1, v).$$

The remaining task is to estimate the difference of dissipations on the right hand side. By a case study we will find for the difference of dissipations the estimates (4.43) and (4.44).

In the first case of either $z_1 = z_0$ or of a negative normal force $\bar{\sigma}(t_2, z_0) \leq 0$ we find for the difference of dissipations on the right hand side the estimate

$$\bar{\Psi}(t_2, z_0, v) - \bar{\Psi}(t_2, z_1, v) \leq 0. \tag{4.43}$$

This is sufficient since the continuous extension of the Assumption (O5), represented by the constant $g_e > 0$, satisfies

$$g_e \leq \bar{\alpha}_e = \min \left\{ \underline{a}(\Delta_e, r_e, v) : v \in \mathbb{S}^1 \right\},$$

see (4.30) and (4.31). The last three inequalities together then prove our result (4.42).

In the second case we assume $\|z_1 - z_0\| > 0$ and $\bar{\sigma}(t_2, z_0) \geq 0$ and we will estimate the difference of dissipations by integrals, see (4.44). Due to $\bar{\sigma}(t_2, z_0) \geq 0$ we find

$$\bar{\Psi}(t_2, z_0, v) - \bar{\Psi}(t_2, z_1, v) = \bar{\sigma}(t_2, z_0)_+ \|\bar{M}(z_0)v\| - \bar{\sigma}(t_2, z_1)_+ \|\bar{M}(z_1)v\|$$
$$\leq (\bar{\sigma}(t_2, z_0) - \bar{\sigma}(t_2, z_1)) \|\bar{M}(z_0)v\| + \bar{\sigma}(t_2, z_1)_+ \left(\|\bar{M}(z_0)v\| - \|\bar{M}(z_1)v\| \right).$$

For $s \in [0,1]$ we define $z(s) := z_1 - s(z_1 - z_0) \in \mathcal{A}_e$ and $m(s) := \|\bar{M}(z(s))v\|$. Since we have $\|v\| = 1$ we find $\|\bar{M}(z(s))v\| > 0$ for all $s \in [0,1]$ by our choice of the set $\mathcal{A}_e$ in (4.29) and we deduce $m \in C^1([0,1],\mathbb{R})$ with

$$-m'(s) = \frac{(\bar{M}(z(s))v)^\top}{\|\bar{M}(z(s))v\|} D\bar{M}(z(s)) \left[v, z_1 - z_0\right].$$

We took here the minus in the presentation since we were seeking for the difference $(z_1 - z_0)$ in the arguments. With the help of the functions z and m we find

$$\bar{\Psi}(t_2, z_0, v) - \bar{\Psi}(t_2, z_1, v)$$
$$\leq \int_0^1 -D\bar{\sigma}(t_2, z(s))(z_1 - z_0)\|\bar{M}(z_0)v\| - \bar{\sigma}(t_2, z_1)_+ \, m'(s) \, \mathrm{d}s.$$

Next we make use of the assumptions $\|z_1 - z_0\| > 0$ and $\|z_2 - z_1\| \geq \|z_1 - z_0\|$. We want to replace the difference $z_1 - z_0$ by $w = \frac{z_1 - z_0}{\|z_1 - z_0\|}$. For this reason we multiply the right hand side of the above inequality with $\frac{\|z_2 - z_1\|}{\|z_1 - z_0\|} > 1$ but first take the modulus of it to make sure that the inequality remains correct. Using all this we conclude

$$\bar{\Psi}(t_2, z_0, v) - \bar{\Psi}(t_2, z_1, v) \tag{4.44}$$
$$\leq \int_0^1 \left| D\bar{\sigma}(t_2, z(s))w\|\bar{M}(z_0)v\| + \bar{\sigma}(t_2, z_1)_+ \frac{(\bar{M}(z(s))v)^\top}{\|\bar{M}(z(s))v\|} DM(z(s)) \left[v, w\right] \right| \mathrm{d}s \|z_2 - z_1\|$$
$$\leq \underline{q}(\Delta_e, r_e, v, w)\|z_2 - z_1\|.$$

Note that the definition of the function $\underline{q}$ in (4.24) was such that exactly this estimate holds. We now join our estimates to obtain

$$\|z_2 - z_1\| \left(\underline{a}(\Delta_e, r_e, v) - \underline{q}(\Delta_e, r_e, v, w)\right)$$
$$\leq (\bar{c}_1 + \bar{c}_2) \max \left\{|\tau_2 - \tau_1|, |t_2 - t_1|\right\}.$$

The functions $\underline{a}$ and $\underline{q}$ still depend on the sliding directions $v, w \in \mathbb{S}^1$ as desired. Next we exploit that by construction we have $z_2 \in \mathcal{K}_\kappa(z_1, z_0)$ or equivalently $v \in \mathcal{K}_\kappa(0, -w)$. Using the definitions (4.25), (4.26) and (4.30) we finally find

$$\mathrm{g}_e \leq G(\Delta_e, r_e, \kappa) \leq \left(\underline{a}(\Delta_e, r_e, v) - \underline{q}(\Delta_e, r_e, v, w)\right)$$

for all $w \in \mathbb{S}^1, v \in \mathcal{K}_\kappa(0, -w)$. The last two estimates together prove our desired estimate (4.42).   ∎

Up to now we have proven two Lipschitz estimates: (4.41) if all solutions z_0, z_1 and z_2 are inside of a tube $\mathcal{T}_\rho$ and (4.42) if the second solution z_2 is in a cone $\mathcal{K}_\kappa$ with respect to the previous solutions z_0 and z_1. In the next lemma we show that the second solution z_2 is automatically in a cone if the solutions are outside of a tube and close to each other.

For the following lemma we define for $\varepsilon > 0$ and $z \in \mathbb{R}^2$ the set

$$\mathcal{M}_\varepsilon(z) := \left\{ y \in \mathbb{R}^2 \ : \ \|\bar{\mathrm{M}}(z)(y-z)\| \leq \varepsilon \right\}. \tag{4.45}$$

Proposition 4.35 (Good changes of direction outside of tube) *We assume* (O0)–(O3) *and* (O5). *Further let the parameter* $\kappa \in (0,1]$ *be given. Then there exists a constant* $c_\kappa > 0$ *such that for all* $\varepsilon > 0$ *the following holds. For the times* $\tau_1, \tau_2, t_1, t_2 \in [T_0, T_0 + \Delta_e]$ *and the value* $z_0 \in \mathcal{A}_e$ *we define first* $z_1 \in \mathcal{A}_e$ *and then* $z_2 \in \mathcal{A}_e$ *by*

$$z_1 \in \operatorname{argmin} \left\{ \bar{\mathcal{E}}_\rho(\tau_1, y) + \bar{\Psi}(t_1, z_0, y - z_0) \ : \ y \in \mathcal{A}_e \right\} \quad \text{and}$$
$$z_2 \in \operatorname{argmin} \left\{ \bar{\mathcal{E}}_\rho(\tau_2, y) + \bar{\Psi}(t_2, z_1, y - z_1) \ : \ y \in \mathcal{A}_e \cap \mathcal{M}_\varepsilon(z_1) \right\}.$$

Now assume $\mathcal{M}_\varepsilon(z_1) \subset \mathcal{A}_e$ *and that the solutions are close the each other and not in a tube, i.e.* $\max \left\{ |\tau_2 - \tau_1|, \|z_2 - z_1\|, \|z_1 - z_0\| \right\} \leq c_\kappa$ *and* $(\tau_j, z_j) \notin \mathcal{T}_{\rho/4}$ *for* $j = 1, 2$. *Then we have*

$$z_2 \in \mathcal{K}_\kappa(z_1, z_0). \tag{4.46}$$

Proof: In the case of $\|z_2 - z_1\| \|z_1 - z_0\| = 0$ the inclusion (4.46) is a direct consequence of the Definition 4.21 of the cone

$$\mathcal{K}_\kappa(z_1, z_0) = \left\{ z \in \mathbb{R}^2 \ : \ \langle z - z_1, z_1 - z_0 \rangle \geq \|z - z_1\| \|z_1 - z_0\| (1 - \kappa) \right\}.$$

Hence, we assume in the following $\|z_2 - z_1\| \|z_2 - z_1\| \neq 0$. In this situation the inclusion (4.46) is equivalent to

$$\left\langle \frac{z_2 - z_1}{\|z_2 - z_1\|}, \frac{z_1 - z_0}{\|z_1 - z_0\|} \right\rangle \geq (1 - \kappa).$$

For $j = 1, 2$ we denote $v_j := \frac{z_j - z_{j-1}}{\|z_j - z_{j-1}\|} \in \mathbb{S}^1$ and calculate due to the Cauchy-Schwarz inequality $\langle v_2, v_1 \rangle = \langle v_2, v_2 - (v_2 - v_1) \rangle \geq 1 - \|v_2 - v_1\|$. Hence, it is sufficient to prove

$$\left\| \frac{z_2 - z_1}{\|z_2 - z_1\|} - \frac{z_1 - z_0}{\|z_1 - z_0\|} \right\| \leq \kappa. \tag{4.47}$$

In the following we calculate explicitly the directions $\frac{z_2 - z_1}{\|z_2 - z_1\|}$ and $\frac{z_1 - z_0}{\|z_1 - z_0\|}$. For $j = 1, 2$ the minimizer z_j satisfies

$$z_j \in \operatorname{argmin} \left\{ \bar{\mathcal{E}}_\rho(\tau_j, y) + \bar{\Psi}(t_j, z_{j-1}, y - z_{j-1}) + \mathcal{X}_{\mathcal{G}_j}(y) \ : \ y \in \mathbb{R}^2 \right\}$$

with the convex sets $\mathcal{G}_1 := \mathcal{A}_e$, $\mathcal{G}_2 := \mathcal{M}_\varepsilon(z_1) \cap \mathcal{A}_e$ and the characteristic functions $\mathcal{X}_{\mathcal{G}_j}(y) := 0$ for $y \in \mathcal{G}_j$ and $\mathcal{X}_{\mathcal{G}_j}(y) := \infty$ else. Due to the convexity of $\mathcal{E}(\tau_j, \cdot)$ on $\mathcal{A}_e$ we use the *Moreau-Rockafellar* Theorem A.3 and find

$$-\mathrm{D}\bar{\mathcal{E}}_\rho(\tau_j, z_j) \in \partial_v \bar{\Psi}(t_j, z_{j-1}, z_j - z_{j-1}) + \partial \mathcal{X}_{\mathcal{G}_j}(z_j) \quad \subset \mathbb{R}^{2^*}.$$

The motivation for deriving the inclusion is that the formula for the subdifferential $\partial_v \bar{\Psi}$ contains the desired direction. Further, the subdifferential coincides with the derivative because we assumed $z_j - z_{j-1} \neq 0$ for the third argument. We omit the lengthy formula and present it in a compromised form, see also Theorem 2.3. We have

$$\partial_v \bar{\Psi}(t_{j-1}, z_{j-1}, z_j - z_{j-1}) = \eta_j \frac{(z_j - z_{j-1})^\top}{\|z_j - z_{j-1}\|} \left(\bar{\mathrm{M}}(z_{j-1})\right)^\top \bar{\mathrm{M}}(z_{j-1}) \quad \in \mathbb{R}^{2^*}$$

for some positive constants $\eta_j > 0$ for $j = 1, 2$. The constants η_j consist basically of the normal force multiplied with some norms.

For the subdifferentials of the characteristic functions we find $\partial \mathcal{X}_{\mathcal{G}_1}(z_1) = 0 \in \mathbb{R}^{2^*}$ since $\mathcal{M}_\varepsilon(z_1) \subset \mathcal{A}_e$ implies $z_1 \in \mathrm{int} \mathcal{A}_e = \mathrm{int} \mathcal{G}_1$. For the second subdifferential $\partial \mathcal{X}_{\mathcal{G}_2}(z_2)$ with $\mathcal{G}_2 = \mathcal{M}_\varepsilon(z_1)$ we find exactly the same form as for the dissipation. Hence, for each $v^* \in \partial \mathcal{X}_{\mathcal{G}_1}(z_2)$ there exists some $\lambda_2 \geq 0$ such that

$$v^* = \lambda_2 \frac{(z_j - z_{j-1})^\top}{\|z_j - z_{j-1}\|} \left(\bar{\mathrm{M}}(z_{j-1})\right)^\top \bar{\mathrm{M}}(z_{j-1}) \quad \in \mathbb{R}^{2^*}.$$

Obviously, we have chosen the set $\mathcal{M}_\varepsilon(z_1) \subset \mathcal{A}_e$ as the minimizing region in the definition of z_2 to get exactly this formula again. Summarizing there exists for $j = 1, 2$ constants $\eta_j > 0$ and $\lambda_j \geq 0$ (take $\lambda_1 = 0$) such that

$$-\mathrm{D}\bar{\mathcal{E}}_\rho(\tau_j, z_j) = (\eta_j + \lambda_j) \frac{(z_j - z_{j-1})^\top}{\|z_j - z_{j-1}\|} \left(\bar{\mathrm{M}}(z_{j-1})\right)^\top \bar{\mathrm{M}}(z_{j-1}) \quad \in \mathbb{R}^{2^*}.$$

This formula motivates the introduction of the direction function

$$F(\tau, x, y) := -\frac{\left(\left(\bar{\mathrm{M}}(y)\right)^\top \bar{\mathrm{M}}(y)\right)^{-1} \mathrm{D}\bar{\mathcal{E}}_\rho^\top(\tau, x)}{\left\| \left(\left(\bar{\mathrm{M}}(y)\right)^\top \bar{\mathrm{M}}(y)\right)^{-1} \mathrm{D}\bar{\mathcal{E}}_\rho^\top(\tau, x) \right\|} \quad \in \mathbb{S}^1 \subset \mathbb{R}^2$$

and with this new notation we find for $j = 1, 2$ our directions described as follows $\frac{z_j - z_{j-1}}{\|z_j - z_{j-1}\|} = F(\tau_j, z_j, z_{j-1})$. Note that the function F is defined and continuous for all $\tau \in [T_0, T_0 + \Delta_e]$ and $x, y \in \mathcal{A}_e$ apart of those pairs (τ, x) for which $\mathrm{D}\bar{\mathcal{E}}_\rho(\tau, x) = 0 \in \mathbb{R}^{2^*}$ holds. These pairs satisfy automatically $x = \bar{z}_f(\tau)$. This motivated the introduction of our tube $\mathcal{T}_{\rho/4}$ in Definition 4.24, because the function $F : \left([T_0, T_0 + \Delta_e] \times \mathcal{A}_e \times \mathcal{A}_e\right) \backslash \left(\mathcal{T}_{\rho/4} \times \mathcal{A}_e\right) \to \mathbb{R}^2$ is uniformly continuous for every choice of $\rho > 0$. Hence there exists a constant $c_\kappa > 0$, independent of ε such that for $j = 1, 2$ and any values $(\tau_j, x_j, y_j) \in \left([T_0, T_0 + \Delta_e] \times \mathcal{A}_e \times \mathcal{A}_e\right) \backslash \left(\mathcal{T}_{\rho/4} \times \mathcal{A}_e\right)$ with $\max\left\{|\tau_2 - \tau_1|, \|x_2 - x_1\|, \|y_2 - y_1\|\right\} \leq c_\kappa$ we find

$$\|F(\tau_2, x_2, y_2) - F(\tau_1, x_1, y_1)\| \leq \kappa. \tag{4.48}$$

But together with the observed equality $F(\tau_j, z_j, z_{j-1}) = \frac{z_j - z_{j-1}}{\|z_j - z_{j-1}\|}$ for $j = 1, 2$ and our assumptions on z_0, z_1, z_2 and τ_1, τ_2 this proves the desired estimate (4.47). And we establish the result.   ∎

The following theorem is the principal result of this section and is the analogon of Theorem 3.19. As a motivation we discuss the theorem in detail. The Assumptions (O1)–(O3) and (O5) are the same as in the final existence Theorem 4.2. The assumption $\alpha > \bar{q}$ assures the strong Assumption (O5*) to hold within the tube. We define the Lipschitz constant

$$c_{\text{lip}} := \max\left\{ \frac{\bar{c}_1 + \bar{c}_2}{\alpha - \bar{q}}, \frac{\bar{c}_1 + \bar{c}_2}{g_e} \right\} > 0 \tag{4.49}$$

with the constants $\bar{c}_1, \bar{c}_2, \alpha, \bar{q}$ and g_e as defined in (4.37)–(4.40) and (4.30). Note that this constant $c_{\text{lip}} > 0$ is independent of the precise choice of the parameter $\rho > 0$. Even if an extended version of the Joint Convexity Assumption (O5) holds within $[T_0, T_0 + \Delta_e] \times \mathcal{A}_e$ we will have to restrict ourself to a shorter time span $\Delta \in (0, \Delta_e]$ defined via

$$\Delta := \min\left\{ \Delta_e, \frac{r_e}{2c_{\text{lip}}} \right\} > 0. \tag{4.50}$$

The choice of the time span is such that we can expect solutions that satisfy the Lipschitz constant c_{lip} and start at $\bar{z}_0$, to remain in the interior of the admissible set $\mathcal{A}_e = \mathcal{B}_{r_e}(\bar{z}_0) \subset \mathbb{R}^2$. Hence these solutions will get into contact with the boundary $\partial \mathcal{A}_e$ and we do not have to take into account artifical constraint forces in the calculus below. The independence of the time span $\Delta > 0$ from the parameter $\rho > 0$ is important since we want to consider later on $\rho_j \to 0$ and choose then always the same time span. Finally the theorem only holds for small fineness of the partitions. This is due to several reasons. For example we have to assure that if a solution is deep inside of the tube, then the next solution is still inside of the tube and vice versa. This will only hold for small step sizes.

Theorem 4.36 (Lipschitz continuity for (IP_ρ)) *Let (O1)–(O3) and (O5) hold. Further we assume $\alpha > \bar{q}$ and define c_{lip} and Δ as in (4.49) and (4.50). Then there exists an upper limit $C_{\text{fine}} > 0$ for the fineness such that for each equi-distant partition Π of $[T_0, T_0 + \Delta]$ satisfying*

$$f_\Pi \leq C_{\text{fine}} \tag{4.51}$$

the solution $(z_k)_{k=0,\ldots,N_\Pi}$ of the corresponding Incremental Problem (IP_ρ), constructed in Theorem 4.30, satisfies for $k = 1, \ldots, N_\Pi$ the uniform Lipschitz estimate

$$\|z_k - z_{k-1}\| \leq c_{\text{lip}}|t_k - t_{k-1}|. \tag{4.52}$$

Proof: The existence of a discrete solution follows from Theorem 4.30. The rest of the proof is carried out by an induction and is based on the previous Propositions 4.31, 4.33 and 4.35.

To be able to apply the results of the Propositions 4.31 and 4.35 we introduce an auxiliary incremental problem. For $z \in \mathbb{R}^2$ and $\varepsilon > 0$ we define the closed and convex set $\mathcal{M}_\varepsilon(z) \subset \mathbb{R}^2$ as in Proposition 4.35 via $\mathcal{M}_\varepsilon(z) := \{y \in \mathbb{R}^2 : \|\bar{\mathrm{M}}(z)(y - z)\| \le \varepsilon\}$. The exact choice of $\varepsilon > 0$ will be specified later.

For $k = 1, \ldots, N_{\Pi}$ we define incrementally the sets $\mathcal{A}_k \subset \mathcal{A}_{\mathrm{e}}$ and the values $\check{z}_k \in \mathcal{A}_{\mathrm{e}}$ by

$$\mathcal{A}_k := \begin{cases} \mathcal{A}_{\mathrm{e}} & \text{if } k = 1, \\ \mathcal{A}_{\mathrm{e}} \cap \mathcal{B}_{\rho/4}(\check{z}_{k-1}) & \text{if } k \ge 2 \text{ and } (t_k, \check{z}_{k-1}) \in \mathcal{T}_{\rho/2}, \\ \mathcal{A}_{\mathrm{e}} \cap \mathcal{M}_\varepsilon(\check{z}_{k-1}) & \text{if } k \ge 2 \text{ and } (t_k, \check{z}_{k-1}) \notin \mathcal{T}_{\rho/2} \end{cases} \quad \text{and}$$

$$\check{z}_k \in \operatorname{argmin}\{\bar{\mathcal{E}}_\rho(t_k, y) + \bar{\Psi}(t_{k-1}, \check{z}_{k-1}, y - \check{z}_{k-1}) : y \in \mathcal{A}_k\}.$$

The existence and uniqueness of a solution $(\check{z}_k)_{k=0,\ldots,N_{\Pi}}$ follows from Proposition 3.10. By induction and using 4.31, 4.33 and 4.35 we now prove for each $k = 1, \ldots, N_{\Pi}$ that

$$\|\check{z}_k - \check{z}_{k-1}\| \le c_{\mathrm{lip}} f_{\Pi} \quad \text{and} \tag{4.53}$$

$$\check{z}_k = z_k \quad \text{hold.} \tag{4.54}$$

For all $k = 1, \ldots, N_{\Pi}$ inequality (4.53) implies $\|\check{z}_k - \check{z}_{k-1}\| \le c_{\mathrm{lip}}|t_k - t_{k-1}|$ since we have equi-distant partitions. Hence the induction proves the desired Lipschitz estimate (4.52).

We start the induction, put $k = 1$ and consider the initial value $\bar{z}_0 \in \mathcal{A}_{\mathrm{e}}$ first. In Lemma 4.27 we have shown that $\bar{z}_0$ satisfies the initial Assumption (3.11) which is equivalent to the minimizing problem

$$\bar{z}_0 \in \operatorname{argmin}\{\bar{\mathcal{E}}_\rho(T_0, y) + \bar{\Psi}(T_0, \bar{z}_0, y - \bar{z}_0) : y \in \mathcal{A}_{\mathrm{e}}\}.$$

By definition we have $T_0 = t_0$ and

$$\check{z}_1 \in \operatorname{argmin}\{\bar{\mathcal{E}}_\rho(t_1, y) + \bar{\Psi}(t_0, \bar{z}_0, y - \bar{z}_0) : y \in \mathcal{A}_{\mathrm{e}}\}.$$

We are now in the setting of Proposition 4.33 with the choice $\mathcal{M}(\bar{z}_0) = \mathbb{R}^2$ and the cone $\mathcal{K}_\kappa(\bar{z}_0, \bar{z}_0) = \mathbb{R}^2$. This simple formula for the cone motivated our choice $\mathcal{A}_1 = \mathcal{A}_{\mathrm{e}}$. Applying Proposition 4.33 and due to $T_0 = t_0$ we then deduce

$$\|\check{z}_1 - \bar{z}_0\| \le \frac{\bar{c}_1 + \bar{c}_2}{g_{\mathrm{e}}}|t_1 - t_0| \le c_{\mathrm{lip}}|t_1 - t_0|.$$

This is the estimate (4.53) for $k = 1$. The states z_1 and $\check{z}_1$ coincide since they are the unique solutions of the identical minimizing problem. This proves the equality (4.54).

For the induction step we fix $k \in \{2, \ldots, N_\Pi\}$ and assume that the estimate (4.53) and the equality (4.54) hold for $j = 1, \ldots, k-1$. Uniqueness implies that $\check{z}_{k-1}$ and z_{k-1} solve the same minimizing problem and we replace the set $\mathcal{A}_{k-1}$ by $\mathcal{A}_e$ in the definition of $\check{z}_{k-1}$. Thus, we find

$$\check{z}_{k-1} \in \operatorname{argmin} \left\{ \bar{\mathcal{E}}_\rho(t_{k-1}, y) + \bar{\Psi}(t_{k-2}, \check{z}_{k-2}, y - \check{z}_{k-2}) : y \in \mathcal{A}_e \right\} \quad \text{and}$$
$$\check{z}_k \in \operatorname{argmin} \left\{ \bar{\mathcal{E}}_\rho(t_k, y) + \bar{\Psi}(t_{k-1}, \check{z}_{k-1}, y - \check{z}_{k-1}) : y \in \mathcal{A}_k \right\}.$$

Depending on the precise definition of the set $\mathcal{A}_k$ we are now formally either in the situation of Proposition 4.31 or Proposition 4.35.
Let us start with the case of $(t_k, \check{z}_{k-1}) \in \mathcal{T}_{\rho/2}$ and thus $\mathcal{A}_k = \mathcal{A}_e \cap \mathcal{B}_{\rho/4}(\check{z}_{k-1})$. But before we apply Proposition 4.31 we qualify for the first time our choice of the upper bound $C_{\text{fine}} > 0$ of the fineness f_Π (see (4.51)). We assume that C_{fine} satisfies $C_{\text{fine}} \leq \frac{\rho}{4} \min \left\{ 1/(2\|\dot{\bar{z}}_f\|_{L^\infty([T_0, T_0 + \Delta_e], \mathbb{R}^2)}), 1/c_{\text{lip}} \right\}$. The first term is such that for $\mathcal{C} := [t_{k-2}, t_k] \times \mathcal{A}_e \cap \mathcal{B}_{\rho/4}(\check{z}_{k-1})$ we find $\mathcal{C} \subset \mathcal{T}_\rho$. To see this we have to prove for $(t, z) \in \mathcal{C}$ the estimate $\|z - \bar{z}_f(t)\| \leq \rho$. We estimate the left hand side via $\|z - \check{z}_{k-1}\| + \|\check{z}_{k-1} - \bar{z}_f(t_k)\| + \|\bar{z}_f(t_k) - \bar{z}_f(t)\| \leq \rho/4 + \rho/2 + \rho/4$. The second term is such that, using for $k-1$ the estimate (4.53), we find $\|\check{z}_{k-1} - \check{z}_{k-2}\| \leq c_{\text{lip}} f_\Pi \leq \frac{\rho}{4}$. Thus we conclude $(t_{k-2}, \check{z}_{k-2}) \in \mathcal{C}$ as required in 4.31. Applying the proposition and then the estimate (4.53) for $k-1$ we find

$$\|\check{z}_k - \check{z}_{k-1}\| \leq \frac{\bar{c}_1 + \bar{c}_2}{\alpha} f_\Pi + \frac{\bar{q}}{\alpha} \|\check{z}_{k-1} - \check{z}_{k-1}\| \leq \left(\frac{\bar{c}_1 + \bar{c}_2}{\alpha} + \frac{\bar{q}}{\alpha} c_{\text{lip}} \right) f_\Pi.$$

We still have to show that $\left(\frac{\bar{c}_1 + \bar{c}_2}{\alpha} + \frac{\bar{q}}{\alpha} c_{\text{lip}} \right) \leq c_{\text{lip}}$ or equivalently $\frac{\bar{c}_1 + \bar{c}_2}{\alpha} \leq \left(1 - \frac{\bar{q}}{\alpha}\right) c_{\text{lip}}$ holds. Exactly here we use the assumption $\alpha > \bar{q}$, see (4.40). We multiply both sides with the factor $\frac{\alpha}{\alpha - \bar{q}} > 0$. Hence, the inequality is equivalent to $\frac{\bar{c}_1 + \bar{c}_2}{\alpha - \bar{q}} \leq c_{\text{lip}}$ which in turn is satisfied by the definition of c_{lip}. This proves the Lipschitz estimate (4.53) for the first case.
We consider now the second case of $(t_k, \check{z}_{k-1}) \notin \mathcal{T}_{\rho/2}$ and $\mathcal{A}_k = \mathcal{M}_\varepsilon(\check{z}_{k-1})$. Since we are outside of some inner tube we want to apply Proposition 4.35 and show that the minimizer $\check{z}_k$ satisfies $\check{z}_k \in \mathcal{K}_\kappa(\check{z}_{k-1}, \check{z}_{k-2})$ and hence our minimizing direction changed only little. This will allow us to apply Proposition 4.33. Before we have to verify all the technical assumptions of Proposition 4.35 and to specify the choice of the parameter $\varepsilon > 0$. Due to the Lipschitz estimate (4.53) for $j = 1, \ldots, k-1$ we conclude $\|\check{z}_{k-1} - \bar{z}_0\| \leq \sum_{j=1}^{k-1} \|\check{z}_j - \check{z}_{j-1}\| \leq N_\Pi c_{\text{lip}} f_\Pi = c_{\text{lip}} \Delta$. Because of our choice $\Delta := \min \left\{ \Delta_e, \frac{r_e}{2c_{\text{lip}}} \right\}$ we find $\check{z}_{k-1} \in \mathcal{B}_{r_e/2}(\bar{z}_0) \subset \mathcal{B}_{r_e}(\bar{z}_0) = \mathcal{A}_e$. Thus to verify the assumption $\mathcal{M}_\varepsilon(\check{z}_{k-1}) \subset \mathcal{A}_e$ of Proposition 4.35 it is sufficient to assure for $r \leq r_e/2$,

$$\mathcal{M}_\varepsilon(\check{z}_{k-1}) = \left\{ y \in \mathbb{R}^2 : \|\bar{M}(\check{z}_{k-1})(y - \check{z}_{k-1})\| \leq \varepsilon \right\} \subset \mathcal{B}_r(\check{z}_{k-1}). \tag{4.55}$$

We postpone this to step (4.56) and the specification of r since a small set $\mathcal{M}_\varepsilon(\breve{z}_{k-1})$ is useful to satisfy also other technical assumptions. The next assumption we have to satisfy is $(t_k, \breve{z}_k) \notin \mathcal{T}_{\rho/4}$. The assumption $(t_{k-1}, \breve{z}_{k-1}) \notin \mathcal{T}_{\rho/4}$ will be checked later. Since we are in the case of $(t_k, \breve{z}_{k-1}) \notin \mathcal{T}_{\rho/2}$ and have $\breve{z}_k \in \mathcal{M}_\varepsilon(\breve{z}_{k-1})$ the assumptions follows again from the inclusion (4.55) if we choose $r \leq \rho/4$. And if we choose $r \leq c_\kappa$ we assure the assumption $\|\breve{z}_k - \breve{z}_{k-1}\| \leq c_\kappa$. All in all we want to satisfy (4.55) with the choice $r = \min\{r_e/2, \rho/4, c_\kappa\}$. For this we define $c_{\min} := \min\{\|\bar{M}(z)v\| : z \in \mathcal{A}_e, v \in \mathbb{S}^1\}$ and deduce $\mathcal{M}_1(0) \subset \mathcal{B}_{c_{\min}}(0)$. Hence, choosing

$$\varepsilon := c_{\min} \min\left\{\frac{r_e}{2}, \frac{\rho}{4}, c_\kappa\right\} \tag{4.56}$$

we establish (4.55). Note for this that c_κ was independent of ε. The remaining technical assumptions of Proposition 4.35 are realized by further qualifying the choice of our upper limit $C_{\text{fine}} > 0$. We derive $(t_{k-1}, \breve{z}_{k-1}) \notin \mathcal{T}_{\rho/4}$ if we choose $C_{\text{fine}} \leq \rho/(4\|\dot{\breve{z}}_f\|_{L^\infty([T_0, T_0+\Delta], \mathbb{R}^2)})$. This follows from the Definition 4.24 of the tube. The last missing assumption $\max\{|t_k - t_{k-1}|, \|\breve{z}_{k-1} - \breve{z}_{k-1}\|\} \leq c_\kappa$ follows from the Lipschitz estimate (4.53) and thus $\max\{|t_k - t_{k-1}|, \|\breve{z}_{k-1} - \breve{z}_{k-1}\|\} \leq \max\{f_{\Pi}, c_{\text{lip}} f_{\Pi}\}$ and the choice $C_{\text{fine}} \leq c_\kappa \min\{1, \frac{1}{c_{\text{lip}}}\}$. Summarizing we assured all assumptions of Proposition 4.35 by the choice (4.56) for $\varepsilon > 0$ and the above preliminary assumption on C_{fine}. Hence, we finally derive

$$\breve{z}_k \in \mathcal{K}_\kappa(\breve{z}_{k-1}, \breve{z}_{k-2}).$$

Adding this additional information to the definition of $\breve{z}_k$ we conclude that we are in the setting of Proposition 4.33. The further assumption $\|\breve{z}_k - \breve{z}_{k-1}\| > \|\breve{z}_{k-1} - \breve{z}_{k-2}\|$ is no severe restriction since the opposite $\|\breve{z}_k - \breve{z}_{k-1}\| \leq \|\breve{z}_{k-1} - \breve{z}_{k-2}\|$ together with the Lipschitz estimate (4.53) for $k-1$ implies immediately (4.53) for k. Hence, applying Proposition 4.33 we find

$$\|\breve{z}_k - \breve{z}_{k-1}\| \leq \frac{\bar{c}_1 + \bar{c}_2}{g_e} f_{\Pi} \leq c_{\text{lip}} f_{\Pi}.$$

This completes the proof of (4.53).

In the second part of the induction step we prove the equality (4.54), i.e. $\breve{z}_k = z_k$. Our aim is to show that z_k and $\breve{z}_k$ are the unique solution of the same minimizing problem. Hence, we want to replace the set $\mathcal{A}_k$ by $\mathcal{A}_e$ in the definition of $\breve{z}_k$. We introduce the constant $c_{\max} := \max\{\|\bar{M}(z)v\| : z \in \mathcal{A}_e, v \in \mathbb{S}^1\}$ and find $\mathcal{B}_{\varepsilon/c_{\max}}(\breve{z}_{k-1}) \subset \mathcal{M}_\varepsilon(\breve{z}_{k-1})$. We make now the last assumption on the upper limit C_{fine} of the fineness f_{Π} and assume

$$C_{\text{fine}} \leq \frac{1}{2c_{\text{lip}}} \min\left\{\frac{\varepsilon}{c_{\max}}, \frac{\rho}{4}\right\}. \tag{4.57}$$

Note that the choice $\varepsilon = c_{\min} \min\left\{\frac{r_e}{2}, \frac{\rho}{4}, c_\kappa\right\}$ was independent of the fineness of partitions. Together with our Lipschitz estimate (4.53) which we have proven in the first part of the induction step inequality (4.57) implies $\breve{z}_k \in \operatorname{int}\mathcal{M}_\varepsilon(\breve{z}_{k-1})$ and $\breve{z}_k \in \operatorname{int}\mathcal{B}_{\rho/4}(\breve{z}_{k-1})$ and hence we can we can replace the set $\mathcal{A}_k$ by $\mathcal{A}_e$ in the definition of $\breve{z}_k$. This shows that $\breve{z}_k$ and z_k are the unique solution of the same problem and thus coincide. We completed now the induction step.

Summarizing the induction we have found that for the choice

$$C_{\text{fine}} = \min\left\{\frac{\rho}{4\|\dot{\breve{z}}_f\|_{L^\infty([T_0,T_0+\Delta],\mathbb{R}^2)}}, c_\kappa, \frac{c_\kappa}{c_{\text{lip}}}, \frac{\varepsilon}{2c_{\text{lip}}c_{\max}}, \frac{\rho}{8c_{\text{lip}}}\right\}$$

the auxiliary solution $(\breve{z}_k)_{k=0,\ldots,N_\Pi}$ and $(z_k)_{k=0,\ldots,N_\Pi}$ coincide. Further the solutions satisfy for all $k = 1, \ldots, N_\Pi$ the estimate

$$\|z_k - z_{k-1}\| \le c_{\text{lip}}f_\Pi = c_{\text{lip}}|t_k - t_{k-1}|.$$

This is the desired Lipschitz estimate (4.53). ∎

The final result of the second part proves the existence of a convergent subsequence of the approximative solutions constructed in the first part. The result is the analog version of the Convergence Theorem 3.20. Both theorems are based on Theorem 3.15, which is a version of the *Arzela-Ascoli* Theorem.

Theorem 4.37 (Existence for auxiliary Contact Problem 4.28) *Let us assume* (O1)–(O3), (O5) *and* $\alpha > \bar{q}$, *see* (4.40). *Then for arbitrary* $\rho > 0$ *there exists a solution of Problem 4.28.*
More precisely, define $\Delta \in (0, \Delta_e]$ *and* $c_{\text{lip}} > 0$ *independent of* $\rho > 0$ *as in* (4.50) *and* (4.49), *then there exists a function* $z_\rho \in W^{1,\infty}([T_0, T_0+\Delta], \mathcal{A}_e)$ *such that* $z_\rho(T_0) = \bar{z}_0$ *and for almost all* $t \in [T_0, T_0 + \Delta]$ *we have*

$$0 \in D\bar{\mathcal{E}}_\rho(t, z_\rho(t)) + \partial_v \bar{\Psi}(t, z_\rho(t), \dot{z}_\rho(t)) + \mathcal{N}_{\mathcal{A}_e}(z_\rho(t)) \quad \subset \mathbb{R}^{3*}. \qquad (\text{DI}_\rho)$$

Further, the solution satisfies for all $t_1, t_2 \in [T_0, T_0+\Delta]$ *the Lipschitz estimate*

$$\|z_\rho(t_2) - z_\rho(t_1)\| \le c_{\text{lip}}|t_2 - t_1|.$$

Proof: We prove existence for the equivalent (S)&(E)-formulation as it can be found in Problem 3.1. The equivalence of the (S)&(E)-formulation with the above differential inclusion (DI_ρ) follows from the convexity of $\bar{\mathcal{E}}_\rho(t, \cdot)$ on $\mathcal{A}_e$ for all $t \in [T_0, T_0 + \Delta_e]$, see Lemma 4.22 and Theorem 2.9 for more details.

The existence for the (S)&(E)-formulation follows from the generic Existence Theorem 3.26. We check its Assumptions (3.2)–(3.6) and (3.31)–(3.34). The Assumptions (3.2)–(3.6) on $\bar{\mathcal{E}}_\rho$ and $\bar{\Psi}$ in Theorem 3.26 follow directly from (O0)–(O3), (O5) and Lemma 4.27. Next a sequence of partitions with $f_{\Pi^{(n)}} \to 0$ was assumed to be given in Theorem 3.26. We proved in the previous Theorem 4.36 that the sequence of discrete solutions $(z_k^{(n)})_{k=1,\ldots,N_{\Pi(n)}}$ of the corresponding incremental problem (IP_ρ) exists and is uniformly Lipschitz continuous with the Lipschitz constant c_{lip} as defined in (4.49). These are the Assumptions (3.31) and (3.32). Further Theorem 3.15 proves the existence of a convergent subsequence $(z^{(n_j)})_{j\in\mathbb{N}} \subset W^{1,\infty}([T_0, T_0+\Delta], \mathcal{A}_e)$ of the sequence of approximative solutions and thus (3.33), i.e.

$$\|z^{(n_j)} - z\|_{C^0([T_0, T_0+\Delta], \mathcal{A}_e)} \to 0 \quad \text{for } j \to \infty.$$

Since $\bar{\Psi}$ is continuous the last Assumption (3.34) is trivial. We now verified all assumptions of the Existence theorem 3.26 and establish our result.   ∎

4.4.3   Solving the contact subproblem

In this section we prove the existence of a solution $\bar{z} \in W^{1,\infty}([T_0, T_0 + \bar{\Delta}], \mathbb{R}^2)$ of the two-dimensional Contact Subproblem 4.9 under the weak Joint Convexity Assumption (O5). For this we make use of solutions z_ρ for simplified Contact Problems 4.28 with increased convexity whose existence we established in the previous section using the weak Convexity Assumption (O5). The principal idea is to fix in these problems the parameter $\alpha > 0$ and to consider a sequence of parameters $(\rho_j)_{j\in\mathbb{N}}$ with $\rho_j \searrow 0$ for $j \to \infty$. Thus, even for fixed $\alpha > 0$, the additional convexity in the simplified Problems 4.28 vanishes in the limit and we will show that the corresponding solutions z_{ρ_j} converge to the two-dimensional solution $\bar{z}$.

Lemma 4.38 (Solving the two-dimensional Contact Problem 4.9)
We assume (O1)–(O3) *and* (O5) *then the two-dimensional Contact Problem 4.9 has a solution, i.e. there exists $\Delta > 0$ and a function $\bar{z} \in W^{1,\infty}([T_0, T_0+\Delta], \mathbb{R}^2)$ satisfying $\bar{z}(T_0) = \bar{z}_0$ and for almost all $t \in [T_0, T_0+\Delta]$ we have*

$$0 \in D\bar{\mathcal{E}}(t, \bar{z}(t)) + \partial_v \bar{\Psi}(t, \bar{z}(t), \dot{\bar{z}}(t)) \quad \subset \mathbb{R}^{2^*}. \tag{$\overline{\text{DI}}$}$$

Proof: In this proof we approximate the Contact Problem 4.9 by solutions z_ρ of the simplified Problem 4.28. Taking $\rho_j \to 0$ and passing to the limit $z_{\rho_j} \to z$ we will strongly use ideas of [MiR07] but for the reader's convenience we present them here again.

We construct the sequence of simplified problems and for this we chose the parameter $\alpha > \bar{q}$, the Lipschitz constant $c_{lip} > 0$ and the time span $\Delta > 0$ as in (4.40), (4.49) and (4.50). Next we chose a sequence of parameters $(\rho_j)_{j\in\mathbb{N}} \in (0, \infty)$ with $\rho_j \to 0$ and denote by $(z_{\rho_j})_{j\in\mathbb{N}} \subset W^{1,\infty}([T_0, T_0+\Delta], \mathcal{A}_e)$ the corresponding sequence of solutions of the simplified Problem 4.28. In the previous Theorem 4.37 we have proven the existence of such solutions. Further we have shown that the solutions are uniformly Lipschitz continuous since the Lipschitz constant $c_{lip} > 0$ defined in (4.49) is independent of the parameter ρ_j, i.e.

$$\|\dot{z}_{\rho_j}\|_{L^\infty([T_0,T_0+\Delta],\mathbb{R}^2)} \leq c_{lip}. \tag{4.58}$$

This uniform Lipschitz continuity implies the existence of a convergent subsequence, still denoted by $(z_{\rho_j})_{j\in\mathbb{N}} \subset W^{1,\infty}([T_0,T_0+\Delta],\mathcal{A}_e)$, and of a limit function $\bar{z} \in W^{1,\infty}([T_0,T_0+\Delta],\mathcal{A}_e)$, see also Theorem 3.15,

$$\|z_{\rho_j} - \bar{z}\|_{C^0([T_0,T_0+\Delta],\mathcal{A}_e)} \to 0 \quad \text{for } j \to \infty. \tag{4.59}$$

Further for general $t \in [T_0, T_0+\Delta]$ and $z \in \mathcal{A}_e = \mathcal{B}_{r_e}(\bar{z}_0)$ we find the uniform convergence

$$\left|\bar{\mathcal{E}}_\rho(t,z) - \bar{\mathcal{E}}(t,z)\right| = \alpha \left|\sum_{j=1}^{2} f_\rho\big((z-\bar{z}_f(t))_j\big) - \sum_{j=1}^{2} f_\rho{}'\big((\bar{z}_0 - \bar{z}_f(T_0))_j\big) z_j\right| \leq \alpha 9 r_e \rho \tag{4.60}$$

and

$$\left|\partial_t\bar{\mathcal{E}}_\rho(t,z) - \partial_t\bar{\mathcal{E}}(t,z)\right| = \alpha \left|\sum_{j=1}^{2} f_\rho{}'\big((z-\bar{z}_f(t))_j\big)(\dot{z}_f)_j(t)\right| \leq \alpha 3\|\dot{z}_f\|_{L^\infty([T_0,T_0+\Delta],\mathcal{B}_{r_e}(\bar{z}_0))}\rho \tag{4.61}$$

For the last two estimates we recall the Definition 4.25 of $\bar{\mathcal{E}}_\rho$ and f_ρ in (4.33). We used that for $y \in \mathbb{R}$ the estimates $|f_\rho(y)| \leq \frac{3}{2}\rho|y|$ and $|f_\rho{}'(y)| \leq \frac{3}{2}\rho$ hold, see also Figure 4.6. The functions z_{ρ_j} solve simplified Problems 4.28. These problems are equivalent to energetic problems since $\mathcal{E}(t,\cdot)$ is convex on $\mathcal{A}_e$ for all $t \in [T_0,T_0+\Delta]$. Thus for all $j \in \mathbb{N}$ and $t \in [T_0,T_0+\Delta]$ we have

$$\bar{\mathcal{E}}_{\rho_j}(t,z_{\rho_j}(t)) \leq \bar{\mathcal{E}}_{\rho_j}(t,y) + \bar{\Psi}(t,z_{\rho_j}(t),y - z_{\rho_j}(t)) \quad \text{for all } y \in \mathcal{A}_e \quad \text{and} \quad (\text{S}_{\rho_j})$$

$$\bar{\mathcal{E}}_{\rho_j}(t,z_{\rho_j}(t)) + \int_{T_0}^{t} \bar{\Psi}(s,z_{\rho_j}(s),\dot{z}_{\rho_j}(s))\,\mathrm{d}s = \bar{\mathcal{E}}_{\rho_j}(T_0,z_{\rho_j}(T_0)) + \int_{T_0}^{t} \partial_s\bar{\mathcal{E}}_{\rho_j}(s,z_{\rho_j}(s))\,\mathrm{d}s. \quad (\text{E}_{\rho_j})$$

We first prove uniform convergence of all terms involving the energy functional. We split $\bar{\mathcal{E}}_{\rho_j}(t,z_{\rho_j}(t)) - \bar{\mathcal{E}}(t,\bar{z}(t)) = \bar{\mathcal{E}}_{\rho_j}(t,z_{\rho_j}(t)) - \bar{\mathcal{E}}(t,z_{\rho_j}(t)) + \bar{\mathcal{E}}(t,z_{\rho_j}(t)) - \bar{\mathcal{E}}(t,\bar{z}(t))$. For the first difference we use (4.60) and for the second difference we use (4.59) together with the continuity (3.2) of $\mathcal{E}$. This proves uniform convergence for the energies. We argue the same way for the difference $\partial_t\bar{\mathcal{E}}_{\rho_j}(t,z_{\rho_j}(t)) - \partial_t\bar{\mathcal{E}}(t,\bar{z}(t))$. Since the dissipation functional is continuous we directly find for all $t \in [T_0,T_0+\Delta]$ the stability, i.e.

$$\bar{\mathcal{E}}(t,\bar{z}(t)) \leq \bar{\mathcal{E}}(t,y) + \bar{\Psi}(t,\bar{z}(t),y - \bar{z}(t)) \quad \text{for all } y \in \mathcal{A}_e. \tag{\bar{\text{S}}}$$

As in Lemma 3.23 we directly deduce from $(\bar{\mathrm{S}})$ the lower energy estimate, i.e.

$$\bar{\mathcal{E}}(t, \bar{z}(t)) + \int_{T_0}^t \bar{\Psi}(\tau, \bar{z}(\tau), \dot{\bar{z}}(\tau)) \, \mathrm{d}\tau \geq \bar{\mathcal{E}}(T_0, \bar{z}(T_0)) + \int_{T_0}^t \partial_\tau \bar{\mathcal{E}}(\tau, \bar{z}(\tau)) \, \mathrm{d}\tau. \qquad (4.62)$$

It remains the integral in (E_{ρ_j}). We first replace the integrand $\bar{\Psi}(t, z_{\rho_j}(t), \dot{z}_{\rho_j}(t))$ by the function $\bar{\Psi}(t, \bar{z}(t), \dot{z}_{\rho_j}(t))$. Since the arguments are uniformly bounded and their difference $(t, z_{\rho_j}(t), \dot{z}_{\rho_j}(t)) - (t, \bar{z}(t), \dot{z}_{\rho_j}(t))$ converges uniformly, see (4.58) and (4.59), we find the continuity of $\bar{\Psi}$ to be sufficient to justify the replacement. Taking the lim inf in (E_{ρ_j}), see also Lemma 3.24 we find for the dissipational integral

$$\int_{T_0}^t \bar{\Psi}(\tau, \bar{z}(\tau), \dot{\bar{z}}(\tau)) \, \mathrm{d}\tau \leq \liminf_{j \to \infty} \int_{T_0}^t \bar{\Psi}(\tau, \bar{z}(\tau), \dot{z}_{\rho_j}(\tau)) \, \mathrm{d}\tau$$

This proves the upper energy estimate, i.e. the opposite of (4.62). Both estimates together prove the energy equality

$$\bar{\mathcal{E}}(t, \bar{z}(t)) + \int_{T_0}^t \bar{\Psi}(\tau, \bar{z}(\tau), \dot{\bar{z}}(\tau)) \, \mathrm{d}\tau = \bar{\mathcal{E}}(T_0, \bar{z}(T_0)) + \int_{T_0}^t \partial_\tau \bar{\mathcal{E}}(\tau, \bar{z}(\tau)) \, \mathrm{d}\tau. \qquad (\bar{\mathrm{E}})$$

Again, we use the convexity of $\mathcal{E}(t, \cdot)$ to derive the equivalence of $(\bar{\mathrm{S}}) \& (\bar{\mathrm{E}})$ and the inclusion

$$0 \in \mathrm{D}\bar{\mathcal{E}}(t, \bar{z}(t)) + \partial_v \bar{\Psi}(t, \bar{z}(t), \dot{\bar{z}}(t)) + \mathcal{N}_{\mathcal{A}_e}(\bar{z}(t)) \quad \subset \mathbb{R}^{2*} \qquad (4.63)$$

to hold for almost all times $t \in [T_0, T_0 + \Delta]$.

The function $\bar{z} \in \mathrm{W}^{1,\infty}([T_0, T_0 + \Delta], \mathcal{A}_e)$ satisfies the Lipschitz constant $c_{\mathrm{lip}} > 0$, too, and we find $\|\bar{z}(t) - \bar{z}_0\| \leq c_{\mathrm{lip}}\Delta$ for all $t \in [T_0, T_0 + \Delta]$. Recalling $\Delta \leq r_e/(2c_{\mathrm{lip}})$, see (4.50), and $\mathcal{A}_e = \mathcal{B}_{r_e}(\bar{z}_0)$ we find $\mathcal{N}_{\mathcal{A}_e}(\bar{z}(t)) = 0$ for all $t \in [T_0, T_0 + \Delta]$. Hence, the time span is chosen short enough to avoid $\bar{z}$ reaching the boundary and thus no constraint forces appear. Thus, (4.63) is equivalent to our desired differential inclusion $(\overline{\mathrm{DI}})$. ∎

Corollary 4.39 (Existence for the Contact Problem 4.8) *If* (O0)–(O3) *and* (O5) *hold. Then Problem 4.8 has a solution.*

Proof: In Lemma 4.10 we have shown the equivalence of the two Problems 4.9 and 4.8. ∎

Corollary 4.40 (Proof of Theorem 4.2) *The above theorem implies Theorem 4.2*

Proof: Let us assume (O0)–(O5). For the *no contact* case of Theorem 4.2 we already established in Corollary 4.5 the existence of a solution. For the *positive* *normal force* case we have shown in Lemma 4.12 that the existence of a contact solution z_c of Problem 4.8 is sufficient to construct a solution. The idea in Lemma 4.12 was to shorten the time span such that the solution remains pressed onto the boundary. For the *contact and no normal force* case of Theorem 4.2 we define a solution candidate by merging the solution of the Contact Problem 4.8, which we established in the previous theorem, with the solution of the Friction-free Problem 4.3. See Definition 4.15 for the precise construction of a solution candidate. In Theorem 4.17 we have finally shown that a solution candidate solves Problem 4.1.

∎

4.5   Examples

We first present two examples of non-existence of a Lipschitz continuous solution by violating the Assumption (O5). These examples can also be found in [ScM07]. From a physical point of view this assumption assures that no sliding direction exists for which the frictional force decreases faster than the elastic force. Otherwise the sliding velocity becomes unbounded in such a direction.

In the examples we restrict ourselves to a two-dimensional setting and choose the admissible set $\mathcal{A} = \{z \in \mathbb{R}^2 : z_2 \geq 0\}$, a purely quadratic energy $\mathcal{E}(t, z) := \frac{1}{2}z^\top \mathrm{H}z - f(t)z$ with constant Hessian matrix $\mathrm{H} \in \mathbb{R}^{2 \times 2}$ and given external forces $f \in \mathbb{C}^2([0, T], \mathbb{R}^{2^*})$. We assume isotropic friction and hence $\mathrm{M}(z) = \begin{pmatrix} \mu(z_1) & 0 \\ 0 & 0 \end{pmatrix}$ with μ being the classical coefficient of friction and put always $Z_0 \in \partial\mathcal{A}$. Consequently the normal force is $\sigma(t, z) = \left(z^\top \mathrm{H} - f(t)\right) \begin{pmatrix} 0 & 1 \end{pmatrix}^\top$ and the dissipation potential turns out to be $\Psi(t, z, v) = \sigma(t, z)\mu(z_1)|v_1|$ for $z_2 = 0$ and $\Psi \equiv 0$ for $z_2 > 0$. Using the subdifferential formulation our problem to solve is

$$-\left(z(t)\right)^\top \mathrm{H} + f(t) \in \begin{pmatrix} \sigma\left(t, z(t)\right)\mu\left(z_1(t)\right)\mathrm{Sign}\left(\dot{z}_1(t)\right) \\ 0 \end{pmatrix}^\top + \begin{pmatrix} 0 \\ \partial\mathcal{X}_{[0,\infty)}\left(z_2(t)\right) \end{pmatrix}^\top \subset \mathbb{R}^{2^*}.$$

(4.64)

Example 4.41 (First example: varying coefficient of friction)
We choose $T_0 = 0, Z_0 = \begin{pmatrix} 1 \\ 0 \end{pmatrix}$ *and* $\mathrm{H} = \begin{pmatrix} \mathrm{H}_{11} & 0 \\ 0 & \mathrm{H}_{22} \end{pmatrix}$ *and make an ansatz of persistent contact* $z_2(t) \equiv 0$ *or* $z(t) = \begin{pmatrix} z_1(t) \\ 0 \end{pmatrix}$. *In fact, for* $z_2(t) \equiv 0$ *the second row in* (4.64) *reads*

$$-\sigma(t, z(t)) = f_2(t) \in (-\infty, 0]$$

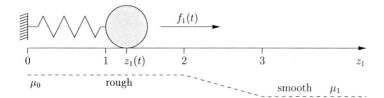

Figure 4.7: First example: varying coefficient of friction. The elastic system consists of a spring with origin in 0 and elongation $z_1(t)$.

and if we choose $f_2(t) = -\sigma_*$ for some constant normal force $\sigma_* > 0$, the above ansatz is justified. It remains to solve the first row in (4.64) that simplifies to

$$-H_{11}z_1(t) + f_1(t) \in \sigma_*\mu(z_1(t))\,Sign(\dot{z}_1(t)).$$

The Convexity Assumption (O5) simplifies to

$$min\left\{v_1 H_{11}v_1 + \sigma_* D\mu\big((Z_0)_1\big)v_1 \,:\, |v_1|=1\right\} = \big(H_{11} - \sigma_*|D\mu\big((Z_0)_1\big)|\big) > 0.$$

To violate this assumption we choose a coefficient μ of friction that depends on z_1. For fixed $\mu_0 > \mu_1 > 0$ we define

$$\mu(z_1) := \begin{cases} \mu_0 & for\ z_1 < 2, \\ \mu_0 + (\mu_1{-}\mu_0)(z_1{-}2) & for\ z_1 \in [2,3], \\ \mu_1 & for\ z_1 > 3. \end{cases}$$

Note that our assumption implies $(\mu_0{-}\mu_1)\sigma_* < H_{11}$. In fact, for the simple force $f_1(t) = H_{11} + at$ with fixed $a > 0$ and $H_{11} > (\mu_0 - \mu_1)\sigma_*$, we obtain the solution

$$z_1(t) = \begin{cases} 1 & for\ t \leq t_1, \\ \frac{a}{H_{11}}t + c_1 & for\ t_1 \leq t \leq t_2, \\ \frac{a}{H_{11}-(\mu_0-\mu_1)\sigma_*}t + c_2 & for\ t_2 \leq t \leq t_3, \\ \frac{a}{H_{11}}t + c_3 & for\ t_3 \leq t, \end{cases}$$

for appropriate values c_1, c_2, c_3 and times t_1, t_2 and t_3, see Figures 4.7 and 4.8. Physically the particle sticks until time t_1. Between time t_1 and time t_2 it slips from 1 to 2 and faces the constant coefficient of friction μ_0. Between time t_2 and t_3 or equivalently while sliding from position 2 to 3 the coefficient of friction μ decreases and the solution $z_1(t)$ is getting closer the friction-free equilibrium $z_f(t) = 1+a/H_{11}t$. This is why we observe here a higher velocity. For $t \geq t_3$ the solution evolves again with the same velocity as z_f but with less distance to the friction-free equilibrium $z_f(t)$ since the coefficient of friction μ_1 is small. Hence, for $H_{11} = (\mu_0{-}\mu_1)\sigma_*$ the *assumption is violated for the values t_2 and $z_1(t_2)$ and regarding our solution we expect a jump to occur at the time $t_2 = t_3$.*

As one sees the possible discontinuity of the solution z_1 has nothing to do with the regularity of μ. We could also choose $\mu \in C^\infty(\mathbb{R}, [0, \infty))$ and provocate a jump by choosing $D\mu$ big enough.

Example 4.42 (Second example: varying normal force) *This second example of non-existence was introduced by Klarbring [Kla90]. As in the first example a jump will occur. This time we assume μ to be constant with $\mu(z_1) = \mu_* > 0$ and the Assumption (O5) reads*

$$min\left\{v^\top Hv + \mu_* D\sigma(T_0, Z_0)v \ : \ v_2 = 0, \|v\| = 1\right\} = \left(H_{11} - \mu_*|H_{12}|\right) > 0.$$

Note that we can violate this assumption by choosing $H_{11} \leq \mu_* H_{12}$. For the initial value we choose again $T_0 = 0$ and $Z_0 = \begin{pmatrix} 1 \\ 0 \end{pmatrix}$. Hence the elastic system is prestressed in normal and tangential direction at time $T_0 = 0$ since we assume for the external tangential force $f_1(t) \equiv 0$. The external normal force is chosen affine in time with $f_2(t) = f_* + at$ and $a > 0$. We have to choose $f_* < 0$ such that the initial state is stable, i.e. the frictional forces are greater than the elastic forces $\mu_* \sigma(0, Z_0) = \mu_* (H_{21} - f_*) \geq H_{11}$. Note that due to $a > 0$ the frictional forces diminish in time and if $H_{11} > \mu_* H_{21}$ holds we have the solution

$$z(t) = \begin{cases} Z_0 & \text{for } t \leq t_1, \\ \begin{pmatrix} \frac{-a}{H_{11} - \mu_* H_{12}} f_2(t) \\ 0 \end{pmatrix} & \text{for } t_1 \leq t \leq t_2, \\ H^{-1} f(t) & \text{for } t_2 \leq t, \end{cases}$$

for appropriate times $0 \leq t_1 < t_2$. Hence, for $t \in [t_1, t_2]$ the body slides from Z_0 to the origin 0. The prestressed system relaxes because of the vanishing friction, while

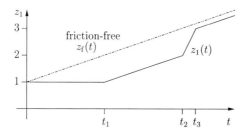

Figure 4.8: First example: solution $z(t) = (z_1(t), 0)^\top$ with large slope in $[t_2, t_3]$ for small $H_{11} - (\mu_0 - \mu_1)\sigma_* > 0$.

for $t > t_2$ we have loss of contact and the position $z(t)$ coincides with the minimizer of $\mathcal{E}$. However, for $H_{11} \leq \mu_* H_{21}$ a jump or spontaneous and complete relaxation occurs at time $t_1 = t_2$ from Z_0 to $H^{-1} f(t_1)$.

An important feature of the idea to consider subproblems was the astonishing observation that the different solutions z_c and z_f coincide at some appropriate time points, even if they solve completely different problems. Hence, one could say that the solutions sometimes do not memorize their evolution. As a concluding remark we now present an example which shows that this important and fruitful feature does not hold if we consider N particles with $N \geq 2$.

Example 4.43 *For $N = 2$ we would have to consider four subproblems and define the solution candidate*

$$z(t) := \begin{cases} z_{cc}(t) & \text{if} \quad \sigma^{(1)}(t, z_{cc}(t)) > 0, \sigma^{(2)}(t, z_{cc}(t)) > 0, \\ z_{cf}(t) & \text{if} \quad \sigma^{(1)}(t, z_{cc}(t)) > 0, \sigma^{(2)}(t, z_{cc}(t)) \leq 0, \\ z_{fc}(t) & \text{if} \quad \sigma^{(1)}(t, z_{cc}(t)) \leq 0, \sigma^{(2)}(t, z_{cc}(t)) > 0, \\ z_{ff}(t) & \text{if} \quad \sigma^{(1)}(t, z_{cc}(t)) \leq 0, \sigma^{(2)}(t, z_{cc}(t)) \leq 0, \end{cases}$$

with z_{cc}, z_{cf}, z_{fc} and z_{ff} denoting the solutions of the constraint-constraint, free-constraint, constraint-free and free-free subproblem. We present an example where the solutions z_{cc} and z_{cf} do not coincide for $\sigma^{(1)}(t, z_{cc}(t)) > 0$ and $\sigma^{(2)}(t, z_{cc}(t)) = 0$ and thus the solution candidate is not continuous.

We have two particles which are in permanent contact, both having a single degree of freedom $z_1, z_2 \in \mathbb{R}$. The physical situation is described in the left picture of Figure 4.9. Hence, we have an external force f which acts only on the second particle and

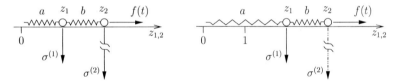

Figure 4.9: Configuration with and without friction at time $t = t_2$

a strong normal and frictional force for this second particle. Without any proof it is clear that if we choose the normal force $\sigma^{(2)}$ big enough than the constrained solution will remain constant, i.e. we will have $\dot{z}_{cc}(t) \equiv 0 \in \mathbb{R}^2$.

For the constraint-free subproblem we neglect the second frictional force and thus the function $z_{cf} = (z_1, z_2)$ has to satisfy

$$0 \in -az_1(t) + b(z_2(t) - z_1(t)) - \mu \sigma^{(1)} \text{Sign}(\dot{z}_1(t)),$$
$$0 \in -b(z_2(t) - z_1(t)) + f(t).$$

The first and second line describes the tangential forces acting on the first and second particle. Let us choose for simplicity $a = b = \mu\sigma^{(1)} = 1$ and the external loading f and normal force $\sigma^{(2)}$ as in Figure 4.10 with $t_* = 1$, $t_1 = 2$ and $t_2 = 4$. We then

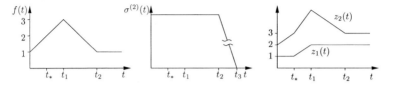

Figure 4.10: Forces $f, \sigma^{(2)}$ and solutions z_1, z_2

find the solution

$$z_1(t) = \begin{cases} 1 & if \quad 0 \leq t \leq t_*, \\ t & if \quad t_* \leq t \leq t_1, \\ 2 & if \quad t_1 \leq t, \end{cases} \qquad z_2(t) = \begin{cases} 2+t & if \quad 0 \leq t \leq t_*, \\ 1+2t & if \quad t_* \leq t \leq t_1, \\ 7-t & if \quad t_1 \leq t \leq t_2, \\ 3 & if \quad t_2 \leq t. \end{cases}$$

which is also represented in the third picture of Figure 4.10. We discuss the solution. Since, we increase the external force f and do not take into account friction for the second particle we find this particle to slide from the beginning on. Due to friction the first particle remains unchanged up to time t_* but then both particles will slide. At time $t = t_1$ we decrease the external force f again. The second friction-free particle reacts immediately but due to friction the first particle sticks. At time $t = t_2$ we returned to our initial external force $f(t_2) = f(0)$ and thus find due to the second inclusion $z_2(0) - z_1(0) = z_2(t_2) - z_1(t_2)$. But due to friction the first particle remembered the evolution of f and we find $z(0) \neq z_1(t_2)$, see the right picture in Figure 4.9. Since at time t_2 the second particle of both solutions z_{cc} and z_{cf} is in a force balance even without friction we can decrease the normal force $\sigma^{(t)}$ for time $t \geq t_2$ such that it disappears at time t_3 and both solutions remain unchanged. Thus at time $t = t_3$ we have $\sigma^{(1)}(t_3, z_{cc}(t_3)) \equiv \sigma^{(1)} > 0$ and for the first time $\sigma^{(2)}(t, z_{cc}(t)) = 0$. We thus change the definition of our solution candidate from $z_{c,c}$ to $z_{c,f}$ and due to $z_{cc}(t_3) \neq z_{cf}(t_3)$ get a jump.

Chapter 5

Making and losing contact (N particles)

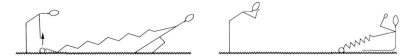

Figure 5.1: Contact vs. non-contact: discontinuous dissipation Ψ

5.1 Overview

In this chapter we allow N particles to make and to lose contact and the frictional forces to vary. But we still assume the boundary to be flat, i.e. $\mathcal{S} = \{y \in \mathbb{R}^3 : y_3 \geq 0\}$ and $\mathcal{A} := \{z \in \mathbb{R}^{3N} : z^{(j)} \in \mathcal{S}, j = 1, \ldots, N\}$. The dissipation functional Ψ is defined by the single dissipation functionals $\Psi^{(j)}$ which are all discontinuous with respect to z due to the possibility of contact and non-contact.

In the previous Example 4.43 we have seen that the elegant trick of Chapter 4 of solving subproblems with continuous dissipations does not work. The example shows that merging the solutions of the subproblems to a solution of the whole problem does not work for $N \geq 2$ particles. This time we do not solve subproblems but construct solutions depending on a parameter $\gamma > 0$. The principal idea is that we introduce new dissipation functionals $\Psi_\gamma^{(j)}$ which are continuous with respect to z at least in situations with small normal forces, i.e. if $\sigma^{(j)}(t, z) < \gamma$ holds. This trick is sufficient since approximate solutions get and lose contact only for small normal forces. We thus avoid the discontinuity of the dissipation.

In Section 5.3 we introduce the new dissipation functionals $\Psi_\gamma^{(j)}$ and define a simplified problem based on these new dissipations. In the following Section 5.4 we then exploit the continuity of the dissipations for small normal forces and construct solutions z_γ. In the last Section 5.5 we show that for $\gamma \to 0$ the functions z_γ converge to a solution of the full problem.

5.2 Assumptions, problem and result

In this section we fix the notation for an N particle problem, introduce the assumptions and present the main result for this chapter, i.e. Theorem 5.2.

A system with $N \in \mathbb{N}$ particles has $3N$ degrees of freedom, our solution vector has the form

$$z = \begin{pmatrix} z_1 \\ z_2 \\ \cdots \\ z_{3N} \end{pmatrix} = \begin{pmatrix} z^{(1)} \\ \cdots \\ z^{(N)} \end{pmatrix} \in \mathbb{R}^{3N} \cong \left(\mathbb{R}^3\right)^N .$$

For $j = 1, \ldots, N$ we denote by $z^{(j)} = \begin{pmatrix} z_{3j-2} & z_{3j-1} & z_{3j} \end{pmatrix}^\top$ the state of the j-th particle. For each single particle we introduce the admissible set

$$\mathcal{S} := \left\{ z \in \mathbb{R}^3 \ : \ z_3 \geq 0 \right\}$$

and claim $z^{(j)} \in \mathcal{S}$ for $j = 1, \ldots, N$. Further, to keep notation simple in the following, it is convenient to introduce for $j = 1, \ldots, N$ the half-spaces

$$\mathcal{A}^{(j)} := \left\{ z \in \mathbb{R}^{3N} \ : \ z_{3j} \geq 0 \right\} .$$

We define now the admissible set $\mathcal{A} \subset \mathbb{R}^{3N}$ for the whole system of N particles in three equivalent ways via

$$\begin{aligned} \mathcal{A} := &\left\{ z \in \mathbb{R}^{3N} \ : \ z_{3j} \geq 0 \quad \text{for all } j = 1, \ldots, N \right\} \\ = &\left\{ z \in \mathbb{R}^{3N} \ : \ z^{(j)} \in \mathcal{S} \quad \text{for all } j = 1, \ldots, N \right\} = \left(\mathcal{S}\right)^N \\ = &\bigcap_{j \in \{1, \ldots, N\}} \mathcal{A}^{(j)} . \end{aligned} \tag{N0}$$

We now introduce the energy functional and assume as usual

$$\mathcal{E} \in \mathrm{C}^2 \left([0, T] \times \mathcal{A}, \mathbb{R}\right) . \tag{N1}$$

We model the dissipation Ψ of the whole system with the help of single dissipation functionals $\Psi^{(j)}$ that describe the dissipation which occurs with respect to each single particle. For $j = 1, \ldots, N$ we denote by

$$\sigma^{(j)}(t, z) := \partial_{z_{3j}} \mathcal{E}(t, z)$$

the normal force with which the j-th particle is pressed onto the obstacle. Further we introduce for each $j = 1, \ldots, N$ a friction matrix $\mathrm{M}^{(j)}$ which is defined with respect to the obstacle surface $\partial \mathcal{S} = \{z \in \mathbb{R}^3 \ : \ z_3 = 0\}$. Of course the roughness of the obstacle surface $\partial \mathcal{S}$ is independent of a single particle but the friction matrix models the interaction between particle and obstacle. Hence, by distinguishing different friction matrices $\mathrm{M}^{(j)}$ for $j = 1, \ldots, N$ we are able to include the possibility that single particles are made of different materials, e.g. rubber, steel, glass, etc. We introduce the normal vector $\nu := \begin{pmatrix} 0 & 0 & -1 \end{pmatrix}^\top$ and assume for $j = 1, \ldots, N$

$$\mathrm{M}^{(j)} \in \mathrm{C}^1\big(\partial \mathcal{S}, \mathbb{R}^{3 \times 3}\big) \quad \text{with} \quad \mathrm{M}^{(j)}(z)\nu(z) = 0 \text{ for all } z \in \partial \mathcal{S}. \tag{N2}$$

Note that $\mathrm{M}^{(j)}$ is evaluated with respect to each single particle $z^{(j)} \in \mathbb{R}^3$ while the energy $\mathcal{E}$ and the normal forces $\sigma^{(j)}$ are defined with respect to the state $z \in \mathbb{R}^{3N}$ of the system.

Next we introduce the dissipational distance of the j-th particle $z^{(j)}$. We recall that $z^{(j)} \in \partial \mathcal{S} = \{y \in \mathbb{R}^3 \ : \ y_3 = 0\}$ is equivalent to $z \in \partial \mathcal{A}^{(j)} = \{z \in \mathbb{R}^{3N} \ : \ z_{3j} = 0\}$. We define a single dissipation $\Psi^{(j)} : [0,T] \times \mathcal{A} \times \mathbb{R}^3 \to [0, \infty)$ via

$$\Psi^{(j)}(t, z, v) := \begin{cases} \sigma^{(j)}(t, z)_+ \, \|\mathrm{M}^{(j)}(z^{(j)})v\| & \text{if } z \in \partial \mathcal{A}^{(j)} \\ 0 & \text{else} \end{cases}, \tag{5.1}$$

with $\sigma^{(j)}(t, z)_+ := \max\{0, \sigma^{(j)}(t, z)\}$ and $\| \cdot \|$ being the usual Euclidian norm. The dissipation functional $\Psi : [0,T] \times \mathcal{A} \times \mathbb{R}^{3N} \to [0, \infty)$ for the whole system is now defined via

$$\Psi(t, z, v) := \sum_{j=1}^{N} \Psi^{(j)}(t, z, v^{(j)}).$$

We now describe the constraint forces $\mathcal{N}_\mathcal{A}$ and we denote by $Z_0 \in \mathcal{A}$ our initial state. Since the admissible set $\mathcal{A} \subset \mathbb{R}^{3N}$ is convex we find for the tangential cone $\mathcal{T}_\mathcal{A}$ the simple formula $\mathcal{T}_\mathcal{A}(Z_0) := \{v \in \mathbb{R}^{3N} \ : \ \lambda v + Z_0 \in \mathcal{A} \text{ for some } \lambda > 0\}$, see Section 2.1.4 for more details. Thus, we define the set of constraint forces via

$$\mathcal{N}_\mathcal{A}(Z_0) := \{v^* \in \mathbb{R}^{3N*} \ : \ v^*v \leq 0 \text{ for all } v \in \mathcal{T}_\mathcal{A}(Z_0)\}.$$

After having redefined the energy functional $\mathcal{E}$ and the dissipational distance Ψ for N particles we can now adopt the same problem formulation (see Problem 4.1) as in the one-particle case.

Problem 5.1 *For given initial time $T_0 \in [0,T)$ and initial state $Z_0 \in \mathcal{A}$ find a positive time span $\Delta \in (0, T-T_0]$ and a solution $z \in \mathrm{W}^{1,\infty}([T_0, T_0+\Delta], \mathcal{A})$ such that the initial condition $z(T_0) = Z_0$ is satisfied and such that for almost all $t \in [T_0, T_0+\Delta]$ the following differential inclusion holds*

$$0 \in \mathrm{D}\mathcal{E}(t, z(t)) + \partial_v \Psi(t, z(t), \dot{z}(t)) + \mathcal{N}_\mathcal{A}(z(t)) \quad \subset \mathbb{R}^{3N*}. \tag{DI}$$

The initial force balance law of elastic, frictional and constraint forces is now described via

$$0 \in D\mathcal{E}(T_0, Z_0) + \partial_v \Psi(T_0, Z_0, 0) + \mathcal{N}_{\mathcal{A}}(Z_0) \quad \subset \mathbb{R}^{3N^*}. \tag{N3}$$

Next we want to make the reader familiar with the structure of the force balance laws (DI) and (N3). We fix some $j = 1, \ldots, N$. The forces that act on the j-th particle $z^{(j)} \in \mathcal{S}$ can be found in the lines $3j-2, 3j-1$ and $3j$. We present these three lines in a rather symbolic way via

$$
\begin{array}{ccccccc}
\begin{pmatrix} 0 \\ 0 \end{pmatrix}^\top & \in & \begin{pmatrix} \partial_{z_{3j-2}} \\ \partial_{z_{3j-1}} \end{pmatrix}^\top \mathcal{E} & + & \partial_{v_1, v_2} \Psi^{(j)} & + & \left\{ \begin{array}{c} 0 \\ 0 \end{array} \right\}^\top & \subset \mathbb{R}^{2^*} \\
0 & \in & \sigma^{(j)} & + & \{0\} & + & f((z^{(j)})_3) & \subset \mathbb{R}^*
\end{array}
\tag{5.2}
$$

and with the set-valued function $f(x) := \begin{cases} \{0\} & \text{if } x > 0, \\ (-\infty, 0] & \text{if } x = 0 \end{cases}$. The first two lines describe the forces in tangential directions, while the third line describes the forces in the normal direction with respect to the boundary of the obstacle $\partial\mathcal{S}$. The frictional forces are represented by the j-th dissipation functional $\Psi^{(j)}$ only. We used $(\partial_v \Psi^{(j)})_3 = \{0\}$ due to $M^{(j)}(z)\nu = 0 \in \mathbb{R}^3$, see Assumption (N2). Hence, the frictional forces act only in tangential directions, which is physically reasonable. The constraint forces show a complementary behavior. They act only in the normal direction. We prove this by deriving an explicit presentation of the constraint forces $\mathcal{N}_{\mathcal{A}}$. Since $\mathcal{A}$ is convex we have $\mathcal{N}_{\mathcal{A}} = \partial \mathcal{X}_{\mathcal{A}}$, see also (2.9). Due to our definition of the admissible set via $\mathcal{A} = \left\{ z \in \mathbb{R}^{3N} : z_{3j} \geq 0 \text{ for } j = 1, \ldots, N \right\}$ an explicit calculus leads us to the formula $\mathcal{N}_{\mathcal{A}}(z) = \begin{pmatrix} 0 & 0 & f((z^{(1)})_3) & \cdots & 0 & 0 & f((z^N)_3) \end{pmatrix} \subset \mathbb{R}^{3N^*}$ with the set-valued function f as defined above. This proves (5.2).

Especially the third line in (5.2) gives a good insight into the problem. Note that normal forces have to be positive $\sigma^{(j)} \geq 0$ and if they are strictly positive, i.e. $\sigma^{(j)}(t, z) > 0$, then the j-th particle has to be in contact, i.e. $(z^{(j)})_3 = 0$. The other way round, if we have no contact, i.e. $(z^{(j)})_3 > 0$, then the third line implies no normal force $\sigma^{(j)}(t, z) = 0$ and hence no friction $\Psi^{(j)}(t, z, \cdot) \equiv 0$. Such simple observations motivated the distinction of three different 'types' of particles. Depending on the initial values T_0 and Z_0 we define the index sets:

$$\mathcal{I} := \{1, \ldots, N\},$$
$$\mathcal{I}_c := \left\{ j \in \mathcal{I} : \sigma^{(j)}(T_0, Z_0) > 0 \right\},$$
$$\mathcal{I}_f := \left\{ j \in \mathcal{I} : (Z_0)_{3j} > 0 \right\} \quad \text{and}$$
$$\mathcal{I}_s := \mathcal{I} \setminus (\mathcal{I}_c \cup \mathcal{I}_f).$$

Due to the initial force balance (N3) we find $\mathcal{I}_c \cap \mathcal{I}_f = \emptyset$ and the sets $\mathcal{I}_c, \mathcal{I}_f$ and $\mathcal{I}_s$ are pairwise disjoint. As usual the index 'c' stands for *contact*, 'f' for *friction-free* and 's' stands for *switching*. Due to (N3) we have for all $j \in \mathcal{I}_c$ that the corresponding particle $(Z_0)^{(j)} \in \mathbb{R}^3$ is in contact, i.e. $(Z_0^{(j)})_3 = 0$. Since it is pressed onto the obstacle with a positive normal force $\sigma^{(j)}(T_0, Z_0) > 0$ we expect it to remain in contact for a short time span. Hence, the index set $\mathcal{I}_c$ represents all particles which are in contact and are expected to remain in contact. Further, the index 'f' denotes *friction-free*. By definition we find $\Psi^{(j)}(T_0, Z_0, \cdot) \equiv 0$ for $j \in \mathcal{I}_f$ and since $(Z_0)_{3j} > 0$ holds we also expect the particle to remain out of contact and thus the index set $\mathcal{I}_f$ denotes all particles which start friction-free and are expected to remain friction-free for a small time span. The index 's' shell represent the word *switching* since for $j \in \mathcal{I}_s = \left\{ j \in \mathcal{I} \ : \ \sigma^{(j)}(T_0, Z_0) = 0, (Z_0)_{3j} = 0 \right\}$ we cannot expect the j-th particle neither to remain in contact nor to remain friction-free. It is possible that the particle switches for arbitrary small time spans infinitely many times between a contact situation with friction and a friction-free situation. See also Example 4.14. For given index set $\mathcal{K} \subset \mathcal{I}$ (the case $\mathcal{K} = \emptyset$ is allowed!) we define the set $\mathbb{S}^{3N-1}_{\mathcal{K}} := \left\{ v \in \mathbb{S}^{3N-1} \ : \ v^j \in \mathcal{T}_{\partial \mathcal{S}}(Z_0), j \in \mathcal{K} \right\} = \left\{ v \in \mathbb{S}^{3N-1} \ : \ v_{3j} = 0, j \in \mathcal{K} \right\}$ and the constants $\alpha_{\mathcal{K}} > 0$ and $q_{\mathcal{K}} > 0$ via

$$\alpha_{\mathcal{K}} := \min \left\{ v^\top \mathrm{H}(T_0, Z_0) v \ : \ v \in \mathbb{S}^{3N-1}_{\mathcal{K}} \right\}, \tag{5.3}$$

$$q_{\mathcal{K}} := \max \Big\{ \sum_{j \in \mathcal{K}} \big| \mathrm{D}\sigma^{(j)}(T_0, Z_0) u \big| \, \| \mathrm{M}^{(j)}(Z_0^{(j)}) v^{(j)} \| \tag{5.4}$$
$$+ \sigma^{(j)}(T_0, Z_0)_+ \| \mathrm{DM}^{(j)}(Z_0^{(j)})[v^{(j)}, u^{(j)}] \| \ : \ u, v \in \mathbb{S}^{3N-1}_{\mathcal{K}} \Big\}.$$

For $\mathcal{K}_1 \subset \mathcal{K}_2$ we find $\mathbb{S}^{3N-1}_{\mathcal{K}_1} \supset \mathbb{S}^{3N-1}_{\mathcal{K}_2}$ and $\alpha_{\mathcal{K}_1} \leq \alpha_{\mathcal{K}_2}$ while no monotonicity holds for $q_{\mathcal{K}}$. With increasing index set $\mathcal{K}$ the number of summands increases but the values of the single summands decrease.

Theorem 5.2 (Existence) *Let us assume* (N0)–(N3)*. For all index sets* $\mathcal{I}_c \subset \mathcal{K} \subset \mathcal{I}_c \cup \mathcal{I}_s$ *we further assume*

$$\alpha_{\mathcal{K}} > q_{\mathcal{K}}. \tag{N4}$$

Then there exists a time span $\Delta > 0$ *and a solution* $z \in \mathrm{W}^{1,\infty}([T_0, T_0 + \Delta], \mathcal{A})$ *of Problem 5.1.*

Proof: We prove existence in Theorem 5.13 below.   ∎

Example 5.3 *The Assumption* (N4) *consists of at most* 2^N *inequalities. In this example we show for the case* $N = 2$ *that none of these assumptions can be omitted or derived from the others. We thus consider a two particle situation with* $z \in \mathbb{R}^6$. *For* $j = 1, 2$ *we assume* $Z_0^{(j)} \in \partial \mathcal{S}$ *and* $\sigma^{(j)}(T_0, Z_0) = 0$, *thus we find* $\mathcal{I}_c = \mathcal{I}_f = \emptyset$ *and* $\mathcal{I}_s = \{1, 2\}$ *and as a consequence* (N4) *consists of four inequalities. For simplicity we assume an isotropic situation of friction such that* $\mathrm{M}^{(j)}(z) =$

$$\begin{pmatrix} \mu(z) & 0 & 0 \\ 0 & \mu(z) & 0 \\ 0 & 0 & 0 \end{pmatrix}.$$ *We have now to check* $\alpha_{\mathcal{K}} > \mathrm{q}_{\mathcal{K}}$ *for all index sets* $\mathcal{I}_{\mathrm{c}} \subset \mathcal{K} \subset \mathcal{I}_{\mathrm{c}} \cup$ $\mathcal{I}_{\mathrm{s}}$, *i.e. for* $\mathcal{K} = \emptyset, \{1\}, \{2\}$ *and* $\{1, 2\}$. *For this we fix the matrix* $\mathrm{H} = \mathrm{H}(T_0, Z_0) \in$ $\mathbb{R}^{6 \times 6}$ *and recall the definitions*

$$\alpha_\emptyset = \min \left\{ \left\langle \begin{pmatrix} H_{11} & H_{12} & H_{13} & H_{14} & H_{15} & H_{16} \\ H_{21} & H_{22} & H_{23} & H_{24} & H_{25} & H_{26} \\ H_{31} & H_{32} & H_{33} & H_{34} & H_{35} & H_{36} \\ H_{41} & H_{42} & H_{43} & H_{44} & H_{45} & H_{46} \\ H_{51} & H_{52} & H_{53} & H_{54} & H_{55} & H_{56} \\ H_{61} & H_{62} & H_{63} & H_{64} & H_{65} & H_{66} \end{pmatrix} \begin{pmatrix} v_1 \\ v_2 \\ v_3 \\ v_4 \\ v_5 \\ v_6 \end{pmatrix}, \begin{pmatrix} v_1 \\ v_2 \\ v_3 \\ v_4 \\ v_5 \\ v_6 \end{pmatrix} \right\rangle : v \in \mathbb{S}^{(5)} \right\},$$

$$\alpha_{\{1\}} = \min \left\{ \left\langle \begin{pmatrix} H_{11} & H_{12} & 0 & H_{14} & H_{15} & H_{16} \\ H_{21} & H_{22} & 0 & H_{24} & H_{25} & H_{26} \\ 0 & 0 & 0 & 0 & 0 & 0 \\ H_{41} & H_{42} & 0 & H_{44} & H_{45} & H_{46} \\ H_{51} & H_{52} & 0 & H_{54} & H_{55} & H_{56} \\ H_{61} & H_{62} & 0 & H_{64} & H_{65} & H_{66} \end{pmatrix} \begin{pmatrix} v_1 \\ v_2 \\ 0 \\ v_4 \\ v_5 \\ v_6 \end{pmatrix}, \begin{pmatrix} v_1 \\ v_2 \\ 0 \\ v_4 \\ v_5 \\ v_6 \end{pmatrix} \right\rangle : v \in \mathbb{S}^{(5)} \right\},$$

$$\alpha_{\{2\}} = \min \left\{ \left\langle \begin{pmatrix} H_{11} & H_{12} & H_{13} & H_{14} & H_{15} & 0 \\ H_{21} & H_{22} & H_{23} & H_{24} & H_{25} & 0 \\ H_{31} & H_{32} & H_{33} & H_{34} & H_{35} & 0 \\ H_{41} & H_{42} & H_{43} & H_{44} & H_{45} & 0 \\ H_{51} & H_{52} & H_{53} & H_{54} & H_{55} & 0 \\ 0 & 0 & 0 & 0 & 0 & 0 \end{pmatrix} \begin{pmatrix} v_1 \\ v_2 \\ v_3 \\ v_4 \\ v_5 \\ 0 \end{pmatrix}, \begin{pmatrix} v_1 \\ v_2 \\ v_3 \\ v_4 \\ v_5 \\ 0 \end{pmatrix} \right\rangle : v \in \mathbb{S}^{(5)} \right\},$$

$$\alpha_{\{1,2\}} = \min \left\{ \left\langle \begin{pmatrix} H_{11} & H_{12} & 0 & H_{14} & H_{15} & 0 \\ H_{21} & H_{22} & 0 & H_{24} & H_{25} & 0 \\ 0 & 0 & 0 & 0 & 0 & 0 \\ H_{41} & H_{42} & 0 & H_{44} & H_{45} & 0 \\ H_{51} & H_{52} & 0 & H_{54} & H_{55} & 0 \\ 0 & 0 & 0 & 0 & 0 & 0 \end{pmatrix} \begin{pmatrix} v_1 \\ v_2 \\ 0 \\ v_4 \\ v_5 \\ 0 \end{pmatrix}, \begin{pmatrix} v_1 \\ v_2 \\ 0 \\ v_4 \\ v_5 \\ 0 \end{pmatrix} \right\rangle : v \in \mathbb{S}^{(5)} \right\}.$$

For the constants $\mathrm{q}_\emptyset, \mathrm{q}_{\{1\}}, \mathrm{q}_{\{2\}}$ *and* $\mathrm{q}_{\{1,2\}}$ *we find due to* $\sigma^{(j)}(T_0, Z_0) = 0$ *the formulas*

$$\mathrm{q}_\emptyset = 0,$$

$$\mathrm{q}_{\{1\}} = \max \left\{ \mu(Z_0^{(1)}) | \left\langle \begin{pmatrix} H_{31} \\ H_{32} \\ 0 \\ H_{34} \\ H_{35} \\ H_{36} \end{pmatrix}, \begin{pmatrix} u_1 \\ u_2 \\ 0 \\ u_4 \\ u_5 \\ u_6 \end{pmatrix} \right\rangle | : u \in \mathbb{S}^{(5)} \right\},$$

$$q_{\{2\}} = \max \left\{ \mu(Z_0^{(2)}) | \langle \begin{pmatrix} H_{61} \\ H_{62} \\ H_{63} \\ H_{64} \\ H_{65} \\ 0 \end{pmatrix}, \begin{pmatrix} u_1 \\ u_2 \\ u_3 \\ u_4 \\ u_5 \\ 0 \end{pmatrix} \rangle | \; : \; u \in \mathbb{S}^{(5)} \right\},$$

$$q_{\{1,2\}} = \max \left\{ \mu(Z_0^{(1)}) | \langle \begin{pmatrix} H_{31} \\ H_{32} \\ 0 \\ H_{34} \\ H_{35} \\ 0 \end{pmatrix}, \begin{pmatrix} u_1 \\ u_2 \\ 0 \\ u_4 \\ u_5 \\ 0 \end{pmatrix} \rangle | + \mu(Z_0^{(2)}) | \langle \begin{pmatrix} H_{61} \\ H_{62} \\ 0 \\ H_{64} \\ H_{65} \\ 0 \end{pmatrix}, \begin{pmatrix} u_1 \\ u_2 \\ 0 \\ u_4 \\ u_5 \\ 0 \end{pmatrix} \rangle | \; : \; u \in \mathbb{S}^{(5)} \right\}.$$

Our main Assumption (N4), *i.e.* $\alpha_{\mathcal{K}} > q_{\mathcal{K}}$ *for all index sets* $\mathcal{I}_c \subset \mathcal{K} \subset \mathcal{I}_c \cup \mathcal{I}_s$
consists of four inequalities:

$$\alpha_\emptyset > 0, \quad \alpha_{\{1\}} > q_{\{1\}}, \quad \alpha_{\{2\}} > q_{\{2\}} \quad and \quad \alpha_{\{1,2\}} > q_{\{1,2\}}.$$

We show that none of these inequalities can be omitted or derived from the other inequalities. To show this we construct examples such that three inequalities hold but not the fourth inequality does not.
Let H *have the structure* $H_{jj} = 1$ *for* $j = 1, \ldots, 6$ *and the only non-zero off-diagonal entry is* $H_{63} = H_{36} = 10$ *and choose* $\mu = 1/20$. *Then* $\alpha_\emptyset > 0$ *does not hold but the other inequalities do with* $\alpha_{\{1\}} = \alpha_{\{2\}} = \alpha_{\{1,2\}} = 1$ *and* $q_{\{1\}} = 1/2$ *and* $q_{\{2\}} = q_{\{1,2\}} = 0$.
Let us next assume $H_{jj} = 10$ *for* $j = 1, \ldots, 6$ *and* $H_{13} = H_{31} = 1$ *to be the only non-zero off-diagonal entry. We then find* $\alpha_\emptyset = \alpha_{\{2\}} = 9$ *and* $\alpha_{\{1\}} = \alpha_{\{1,2\}} = 10$. *Further we have* $q_\emptyset = q_{\{2\}} = q_{\{1,2\}} = 0$. *But choosing* $\mu = 20$ *we get* $q_{\{1\}} = 20$ *and thus* $\alpha_{\{1\}} = 9 < 20 = q_{\{1\}}$ *while the other inequalities hold. For symmetry reasons this also implies that we cannot derive the third inequality* $\alpha_{\{2\}} > q_{\{2\}}$ *from the others.*
As a last example we choose $H_{jj} = 10$ *for* $j = 1, \ldots, 6$ *and* $H_{13} = H_{31} = H_{16} = H_{61} = 1$ *and* $H_{ij} = 0$ *else. We then find* $\alpha_\emptyset = 10 - \sqrt{2}$, $\alpha_{\{1\}} = \alpha_{\{2\}} = 9$ *and* $\alpha_{\{1,2\}} = 10$. *Choosing* $\mu = 6$ *we find* $q_\emptyset = 0$ *and* $q_{\{1\}} = q_{\{2\}} = 6$, *hence the first three inequalities hold. The fourth does not hold since we have* $q_{\{1,2\}} = \mu H_{31} + \mu H_{61} = 12 > 10 = \alpha_{\{1,2\}}$.

5.3   Decreasing the friction: a simplified problem

In this section we present a simplified problem, see Problem 5.4. All simplifications help to avoid the discontinuity of the dissipation functional Ψ. We present three kinds of simplifications but before we start by extending the Assumption (N4) to some neighborhood of the initial values (T_0, Z_0). The index 'e' stands for *extended*. The parameter $l_e > 0$ denotes half of the edge length of the cuboid

$$\mathcal{Q}_{l_e}(Z_0) := \left\{ z \in \mathbb{R}^{3N} \ : \ |(z - Z_0)_n| \leq l_e, n = 1, \dots, 3N \right\}.$$

Further we introduce for a given index set $\mathcal{K} \subset \mathcal{I}$ the convex set $\mathcal{A}_{\mathcal{K},e} := \mathcal{A} \cap \left(\bigcap_{j \in \mathcal{K}} \partial \mathcal{A}^{(j)} \right) \cap \mathcal{Q}_{l_e}(Z_0)$. Note that due to our choice of $\mathcal{Q}_{l_e}$ we can describe the set $\mathcal{A}_{\mathcal{K},e} \subset \mathbb{R}^{3N}$ as an intersection of half-spaces and hyper planes. This will make the formula for the constraint forces easier. In Figure 5.2 we present an elastic system

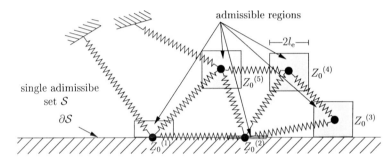

Figure 5.2: Admissible regions for $z \in \mathcal{A}_{\mathcal{K},e} \subset \mathbb{R}^{3N}$ with $\mathcal{K} = \{2\}$ and $N = 5$

in its initial configuration $Z_0 \in \mathbb{R}^{3N}$. If $z \in \mathcal{A}_{\mathcal{K},e}$ holds then each single particle $z^{(j)} \in \mathcal{S} \subset \mathbb{R}^3$ is constrained to its grey box or plane. Since $\sigma^{(1)}(T_0, Z_0) = 0$ holds we free the particle $Z_0(1)$ and allow it to leave the boundary. Due to $\sigma^{(2)}(T_0, Z_0) > 0$ we decide to restrict the second particle to the obstacle. We denote the extended time span by $\Delta_e \in (0, T - T_0]$ and introduce the cylinder $\mathcal{C}_{\mathcal{K},e} := [T_0, T_0 + \Delta_e] \times \mathcal{A}_{\mathcal{K},e}$. The following constants are extended versions of the constants $\alpha_{\mathcal{K}} > 0$ and $q_{\mathcal{K}} > 0$ defined in (5.3) and (5.4). They coincide if we choose $\Delta_e = l_e = 0$.

$$\alpha_{\mathcal{K},e} := \min \left\{ v^\top H(t, z) v \ : \ v \in \mathbb{S}_{\mathcal{K}}^{3N-1}, (t, z) \in \mathcal{C}_{\mathcal{K},e} \right\}, \tag{5.5}$$

$$q_{\mathcal{K},e} := \max \Big\{ \sum_{j \in \mathcal{K}} \left\| D\sigma^{(j)} u \right\|_{L^\infty(\mathcal{C}_{\mathcal{K},e}, \mathbb{R})} \left\| M^{(j)}(\cdot^{(j)}) v^{(j)} \right\|_{L^\infty(\mathcal{A}_{\mathcal{K},e}, \mathbb{R}^3)} \tag{5.6}$$

$$+ \left\| \sigma_+^{(j)} \right\|_{L^\infty(\mathcal{C}_{\mathcal{K},e}, \mathbb{R})} \left\| DM^{(j)}(\cdot^{(j)})[v^{(j)}, u^{(j)}] \right\|_{L^\infty(\mathcal{A}_{\mathcal{K},e}, \mathbb{R}^3)} \ : \ u, v \in \mathbb{S}_{\mathcal{K}}^{3N-1} \Big\}.$$

Due to the regularity of $\mathcal{E}$ and $M^{(j)}$, see (N1) and (N2), and Assumption (N4) there exist $l_e > 0$ and $\Delta_e > 0$ such that for all $\mathcal{K} \subset \mathcal{I}$ with $\mathcal{I}_c \subset \mathcal{K} \subset \mathcal{I}_c \cup \mathcal{I}_s$ we find

$$\alpha_{\mathcal{K},e} > q_{\mathcal{K},e}. \tag{5.7}$$

We now chose $\Delta_{\mathrm{e}} \in (0, T - T_0]$ and $l_{\mathrm{e}} > 0$ such that (5.7) holds and fix this pair $(\Delta_{\mathrm{e}}, l_{\mathrm{e}})$ for the rest of this N *particles* chapter.

We start with the first simplification. For the rest of this chapter we denote

$$\mathcal{A}_{\mathrm{e}} := \mathcal{A} \cap \left(\bigcap_{j \in \mathcal{I}_c} \partial \mathcal{A}^{(j)} \right) \cap \mathcal{Q}_{l_{\mathrm{e}}}(Z_0),$$

and assume $z \in \mathcal{A}_{\mathrm{e}} = \mathcal{A}_{\mathrm{I}_c,\mathrm{e}}$. Note that for $j \in \mathcal{I}_c$ the single dissipation functional $\Psi^{(j)}$ is continuous for z restricted to $z \in \mathcal{A}_{\mathrm{e}}$. This idea corresponds to the procedure we used to solve the Contact Subproblem in the Sections 4.3.2 and 4.4.

For the second simplification we do not include the single dissipation $\Psi^{(j)}$ for all $j \in \mathcal{I}_f$, i.e. we put it zero. See the definition (5.9) below. Since we assume $z \in \mathcal{A}_{\mathrm{e}}$ all *free* particles, i.e. $z^{(j)}$ with $j \in \mathcal{I}_f$, are not restricted to the boundary. Hence, we treat this particles similar to the single particle of the Friction-free Subproblem in Section 4.3.1.

The new idea of this chapter is linked with the third and last simplification and the question how to deal with those particles which are neither pressed onto the boundary nor are out of contact, hence all particles $z^{(j)}$ with $j \in \mathcal{I}_s$. We should neither restrict these particles to the boundary nor should we neglect their dissipations. Hence, we have to deal with the discontinuity of the dissipation due to making and losing contact. But one might ask why an expected physical solution does not face this discontinuity? The answer is intuitively simple. A physical particle only leaves the boundary if its normal force vanishes and it is no longer pressed onto the boundary. But zero normal force implies zero dissipation. Hence, in the moment of leaving the boundary the dissipation functional passes from zero to zero and no discontinuity appears. Unfortunately in general our discrete approximative solutions do not capture this critical moment. Thus it is possible that the j-th particle $z_k^{(j)}$ of a discrete solution z_k is pressed onto the boundary with positive normal force while for the next solution the particle left the boundary, i.e.

$$z_k^{(j)} \in \partial \mathcal{S} \quad \text{with } \sigma^{(j)}(t_k, z_k) > 0 \quad (\text{, hence } \Psi^{(j)}(t_k, z_k, \cdot) \not\equiv 0$$
$$z_{k+1}^{(j)} \in \mathrm{int}\mathcal{S} \quad \text{and hence } \Psi^{(j)}(t_{k+1}, z_{k+1}, \cdot) \equiv 0.$$

In such situations the discontinuity of $\Psi^{(j)}$ matters and we are not able to prove Lipschitz continuity for z_k and z_{k+1} under reasonable assumptions.

(The first idea we had to solve this difficulty was to plug in an additional time point $t_* \in (t_{k-1}, t_k)$ such that the corresponding discrete solution captures the critical moment, i.e. $z_*^{(j)} \in \partial \mathcal{S}$ and $\sigma^{(j)}(t_*, z_*) = 0$. But it shows that infinitely many of such critical time points can occur within the time step (t_{k-1}, t_k), see Example 4.14. This led us to severe technical problems.)

But since the normal forces of the discrete solution are small before and after losing and making contact we modify the dissipation functional such that it is continuous for small normal forces. For this we introduce a parameter $\gamma > 0$ and define for $j \in \mathcal{I}_s$ the single functional via

$$\Psi_\gamma^{(j)} : [0,T] \times \mathcal{A} \times \mathbb{R}^3 \to [0,\infty)$$
$$(t,z,v) \mapsto \begin{cases} \left(\sigma^{(j)}(t,z) - \gamma\right)_+ \|\mathrm{M}^{(j)}(z^{(j)})v\| & \text{if } z_{3j} = 0, \\ 0 & \text{if } z_{3j} > 0. \end{cases} \quad (5.8)$$

with $\left(\sigma(t,z) - \gamma\right)_+ := \max\{0, \sigma(t,z) - \gamma\}$ and $\|\cdot\|$ being the usual euclidian norm. See also Figure 1.9. The motivation for this new dissipation functional can be understood better if we reformulate the presentation of $\Psi_\gamma^{(j)}$ in a more convenient form, i.e.

$$\Psi_\gamma^{(j)}(t,z,v) = \begin{cases} \left(\sigma^{(j)}(t,z) - \gamma\right) \|\mathrm{M}^{(j)}(z^{(j)})v\| & \text{if } z_{3j} = 0, \sigma^{(j)}(t,z) \geq \gamma \\ 0 & \text{if } z_{3j} = 0, \sigma^{(j)}(t,z) < \gamma \\ 0 & \text{if } z_{3j} > 0. \end{cases}$$

This dissipation is still not continuous if we pass directly from the first to the third line. But our discrete solution will not change from the first to the third line or vice versa without passing through a state of small normal forces and hence line two which corresponds to the critical situation. From a more abstract point of view we separated the non-smoothness of $\Psi^{(j)}$ and $\mathcal{N}_{\mathcal{A}^{(j)}}$. We now have $\Psi_\gamma^{(j)}$ to be non-smooth for $\sigma^{(j)} = \gamma$ and $\mathcal{N}_{\mathcal{A}^{(j)}}$ for $z_{3j} = 0$. This will be the key argument in Theorem 5.10 where we prove Lipschitz continuity of discrete solutions with a Lipschitz constant independent of γ. We finally present our modified dissipation functional $\Psi_\gamma : [0,T] \times \mathcal{A} \times \mathbb{R}^{3N} \to [0,\infty)$ with

$$\Psi_\gamma(t,z,v) := \sum_{j \in \mathcal{I}_c} \Psi^{(j)}(t,z,v^{(j)}) + \sum_{j \in \mathcal{I}_s} \Psi_\gamma^{(j)}(t,z,v^{(j)})). \quad (5.9)$$

Problem 5.4 *Find a function* $z_\gamma \in \mathrm{W}^{1,\infty}([T_0, T_0 + \Delta_e], \mathcal{A}_e)$ *such that the initial condition* $z_\gamma(T_0) = Z_0$ *is satisfied and such that for all times* $t \in [T_0, T_0 + \Delta_e]$ *we have*

$$\mathcal{E}(t, z_\gamma(t)) \leq \mathcal{E}(t,y) + \Psi_\gamma(t, z_\gamma(t), y - z_\gamma(t)) \quad \text{for all } y \in \mathcal{A}_e \quad \text{and} \quad (\mathrm{S})$$

$$\mathcal{E}(t, z_\gamma(t)) + \int_{T_0}^t \Psi_\gamma(s, z_\gamma(s), \dot{z}_\gamma(s)) \,\mathrm{d}s = \mathcal{E}(T_0, z_\gamma(T_0)) + \int_{T_0}^t \partial_s \mathcal{E}(s, z_\gamma(s)) \,\mathrm{d}s. \quad (\mathrm{E})$$

5.4 Solving simplified problems

In this section we construct a solution of Problem 5.4 following the scheme of the basic mathematical strategy. We cannot apply the basic strategy one by one, since the dissipation Ψ_γ is not continuous with respect to the state $z \in \mathcal{A}_e$. But we will reuse big parts of it. The existence result is established in Theorem 5.12.

We start applying the basic strategy and verify as many of its Assumptions (3.1)–(3.11) as possible. By definition the set $\mathcal{A}_e \subset \mathbb{R}^{3N}$ is convex and closed, this is (3.1). The regularity Assumption (3.2) on $\mathcal{E}$ follows from its definition. Next we want to show the α-uniform convexity of $\mathcal{E}$. Let us choose $\alpha = \alpha_{\mathcal{I}_c,e}$ then the Assumption (5.7) implies $\alpha > 0$ and by the definition of $\alpha_{\mathcal{I}_c,e}$ we find $\mathcal{E}(t,\cdot)$ to be α-uniformly convex on $\mathcal{A}_e$ for all $t \in [T_0, T_0 + \Delta_e]$. This is (3.3). The three Assumptions (3.4)–(3.6) on Ψ_γ are a direct consequence of its definition. Note that (3.4) holds only in the weak version, i.e. $\Psi(\cdot, z, \cdot) \in \mathrm{C}([T_0, T_0+\Delta], \mathbb{R}^{3N})$ for all $z \in \mathcal{A}_e$. We define the cylinder $\mathcal{C}_e := [T_0, T_0 + \Delta_e] \times \mathcal{A}_e$ and introduce the constants

$$c_1 := \frac{1}{2} \left\| \partial_t \mathrm{D}\mathcal{E} \right\|_{\mathrm{L}^\infty(\mathcal{C}_e, \mathbb{R}^{3N*})} \quad \text{and}$$

$$c_2 := \sum_{j \in \mathcal{I}_c \cup \mathcal{I}_s} \left\| \partial_t \sigma^{(j)} \right\|_{\mathrm{L}^\infty(\mathcal{C}_e, \mathbb{R})} \left\| \mathrm{M}^{(j)} \right\|_{\mathrm{L}^\infty(\partial \mathcal{S} \cap \mathcal{Q}_{l_e}(Z_0^{(j)}), \mathbb{R}^{3\times3})}.$$

Note that both constants are independent of the parameter $\gamma > 0$. This is important since we are seeking for a Lipschitz constant independent of γ and the constants c_1 and c_2 will be part of it. We have chosen the constants $c_1, c_2 > 0$ such that we find

$$\frac{1}{2} \left\| \partial_t \mathrm{D}\mathcal{E}(t, z) \right\| \le c_1 \quad \text{for all } (t, z) \in \mathcal{C}_e \quad \text{and}$$

$$|\Psi_\gamma(t_2, z, v) - \Psi_\gamma(t_1, z, v)| \le c_2 |t_2 - t_1| \quad \text{for all } (t_k, z) \in \mathcal{C}_e, k = 1, 2 \text{ and } v \in \mathbb{S}^{3N-1}_{\mathcal{I} \setminus \mathcal{I}_f}.$$

These are the Assumptions (3.7) and (3.8). We miss the Lipschitz Continuity Assumption (3.9) for $\Psi_\gamma(t, \cdot, v)$ and the corresponding Convexity Assumption (3.10). Due to the convexity of $\mathcal{E}$ the initial Assumption (3.11) is equivalent to the inclusion

$$0 \in \mathrm{D}\mathcal{E}(T_0, Z_0) + \partial_v \Psi_\gamma(T_0, Z_0, 0) + \mathcal{N}_{\mathcal{A}_e}(Z_0) \quad \subset \mathbb{R}^{3N*}.$$

This force balance follows from (N3), i.e. $0 \in \mathrm{D}\mathcal{E}(T_0, Z_0) + \partial_v \Psi(T_0, Z_0, 0) + \mathcal{N}_{\mathcal{A}}(Z_0)$. The frictional forces coincide since the same single dissipation functionals are active for the initial values, we find $\Psi_\gamma(T_0, Z_0, \cdot) \equiv \sum_{j \in \mathcal{I}_c} \Psi^{(j)}(T_0, Z_0, \cdot) \equiv \Psi(T_0, Z_0, \cdot)$. Further, we have $\mathcal{N}_{\mathcal{A}}(Z_0) \subset \mathcal{N}_{\mathcal{A}_e}(Z_0)$ due to the definition of $\mathcal{A}_e$. Summarizing we have shown the following result:

Lemma 5.5 *The Assumptions* (N0)–(N4) *imply that the set* $\mathcal{A}_e \subset \mathbb{R}^{3N}$ *and the functionals* $\mathcal{E}$ *and* Ψ_γ *satisfy the Assumptions* (3.1)–(3.8) *and* (3.11) *of the basic mathematical strategy.*

The missing Lipschitz Continuity Assumptions (3.9) and Convexity Assumption (3.10) hold on restricted sets.

Lemma 5.6 *We assume* $\mathcal{I}_c \subset \mathcal{K} \subset \mathcal{I}_c \cup \mathcal{I}_s$ *to be given and define the admissible set* $\mathcal{A}_{\mathcal{K}} := \mathcal{A}_e \cap \left(\bigcap_{j \in \mathcal{K}} \partial \mathcal{A}^{(j)} \right)$, *the dissipation functional* $\Psi_{\mathcal{K}} := \sum_{j \in \mathcal{I}_c} \Psi^{(j)} + \sum_{j \in \mathcal{K} \setminus \mathcal{I}_c} \Psi_\gamma^{(j)}$ *and the constant* $\alpha_{\mathcal{K},e}, q_{\mathcal{K},e} > 0$ *as in* (5.5), (5.6).

Then, for all $t \in [T_0, T_0 + \Delta_e]$ and $v \in \mathbb{R}^{3N}$ the function $\mathcal{E}(t, \cdot)$ is $\alpha_{\mathcal{K},e}$-uniformly convex and the function $\Psi_{\mathcal{K}}(t, \cdot, v)$ has the Lipschitz constant $q_{\mathcal{K},e}$ on $\mathcal{A}_{\mathcal{K}}$. Further the convexity Assumption $\alpha_{\mathcal{K},e} > q_{\mathcal{K},e}$ holds.

Proof: The properties follow directly from the constructions of $\alpha_{\mathcal{K},e}, q_{\mathcal{K},e} > 0$. We find for the decreased normal force $D(\sigma^{(j)} - \gamma) = D\sigma^{(j)}$ and $\|(\sigma^{(j)} - \gamma)_+\| \leq \|\sigma^{(j)}_+\|$. This shows why $q_{\mathcal{K},e}$ could be chosen independent of $\gamma > 0$. ∎

We start the basic strategy and define the incremental problem and the approximative solutions that correspond to the above Problem 5.4.

Definition 5.7 (Incremental Problem and approximative solutions)
For the given initial value $z_0 = Z_0 \in \mathcal{A}_e$, time span $\Delta \in (0, \Delta_e]$ and a partition $\Pi : T_0 = t_0 < t_1 < \cdots < t_{N_\Pi} = T_0 + \Delta$ the Incremental Problem *consists in finding incrementally $z_1, z_2, z_3, \ldots, z_{N_\Pi}$ such that for $k = 1, \ldots, N_\Pi$*

$$z_k \in \operatorname{argmin}\left\{ \mathcal{E}(t_k, y) + \Psi_\gamma(t_{k-1}, z_{k-1}, y - z_{k-1}) \ : \ y \in \mathcal{A}_e \right\}. \qquad (\text{IP}_\gamma)$$

holds.
For a sequence of equi-distant partitions $(\Pi^{(n)})_{n \in \mathbb{N}}$ of $[T_0, T_0 + \Delta]$, i.e.

$$\Pi^{(n)} : T_0 = t_0^{(n)} < t_1^{(n)} < \cdots < t_{N_{\Pi^{(n)}}} = T_0 + \Delta$$

we define the sequence of approximative solutions *$(z_\gamma^{(n)})_{n \in \mathbb{N}} \subset W^{1,\infty}([T_0, T_0 + \Delta], \mathcal{A}_e)$ as the piecewise linear interpolation between the values $(z_k^{(n)})_{k=1,\ldots,N_{\Pi^{(n)}}}$ of the solution of the corresponding Incremental Problem* (IP_γ), *i.e.*

$$z_\gamma^{(n)}(t) := z_{k-1}^{(n)} + \frac{t - t_{k-1}^{(n)}}{t_k^{(n)} - t_{k-1}^{(n)}}(z_k^{(n)} - z_{k-1}^{(n)}) \quad \text{for } t \in [t_{k-1}^{(n)}, t_k^{(n)}] \text{ and } k = 1, \ldots, N_{\Pi^{(n)}}.$$

Theorem 5.8 (Existence and uniqueness of approximative solutions)
We assume (N0)–(N4). *Then for any time span $\Delta \in (0, \Delta_e]$ and any equi-distant sequence of partitions $(\Pi^{(n)})_{n \in \mathbb{N}}$ of $[T_0, T_0 + \Delta]$ the corresponding sequence of approximative solutions $(z_\gamma^{(n)})_{n \in \mathbb{N}} \subset W^{1,\infty}([T_0, T_0 + \Delta], \mathcal{A}_e)$ of (IP_γ) exists and is unique.*

Proof: The existence and uniqueness of approximative solutions was already proven in Lemma 3.12 under the Assumptions (3.1)–(3.6) on $\mathcal{A}_e, \mathcal{E}$ and Ψ_γ and we have shown in Lemma 5.5 that (N0)–(N4) imply these assumptions. ∎

The following Lemma is only an auxiliary but technically useful result. It will be used in the crucial Theorem 5.10 several times.

Lemma 5.9 *We assume* (N0)–(N4). *Further let* $t_0, t \in [T_0, T_0 + \Delta_e]$, $z_0 \in \mathcal{A}_e$ *and* $\mathcal{I}_* \subset \mathcal{I}_s$ *be given. We assume* $j_* \in \mathcal{I}_*$. *Then the following two situations are equivalent*

$$z \in \operatorname{argmin}\left\{ \mathcal{E}(t, y) + \Psi_\gamma(t_0, z_0, y - z_0) \; : \; y \in \mathcal{A}_e \cap \bigcap_{j \in \mathcal{I}_*} \partial \mathcal{A}^{(j)} \right\}, \sigma^{(j_*)}(t, z) > 0 \; and$$

(5.10)

$$z \in \operatorname{argmin}\left\{ \mathcal{E}(t, y) + \Psi_\gamma(t_0, z_0, y - z_0) \; : \; y \in \mathcal{A}_e \cap \bigcap_{j \in \mathcal{I}_* \setminus \{j_*\}} \partial \mathcal{A}^{(j)} \right\}, \sigma^{(j_*)}(t, z) > 0.$$

(5.11)

Proof: Let us first assume the second situation (5.11) to be satisfied. Our aim is to show $z \in \partial \mathcal{A}^{(j_*)}$, see (5.14) below. We define the cuboid $\mathcal{Q} := \mathcal{A}_e \cap \bigcap_{j \in \mathcal{I}_* \setminus \{j_*\}} \partial \mathcal{A}^{(j)}$. We reformulate the inclusion (5.11) as a force balance law using the *Moreau-Rockafellar* Theorem and conclude

$$0 \in \mathrm{D}\mathcal{E}(t, z) + \partial_v\left(\Psi_\gamma(t_0, z_0, \cdot - z_0)\right)(z) + \partial \mathcal{X}_{\mathcal{Q}}(z) \quad \subset \mathbb{R}^{3N^*}.$$

(5.12)

We consider the $3j_*$-th column which describes physically the normal forces for the j_*-th particle. By definition we find $(\mathrm{D}\mathcal{E})_{3j_*} = \sigma^{(j_*)}$. Further we have $(\partial_v \Psi_\gamma)_{3j_*} = 0 \in \mathbb{R}^*$ since the frictional forces act only in tangential directions, see also Remark 2.4 and Theorem 2.3 for the precise calculus of $\partial_v \Psi_\gamma$. To calculate the constraint force $(\partial \mathcal{X}_{\mathcal{Q}})_{3j_*}$ we describe the above cuboid as an intersection of stripes, i.e. $\mathcal{Q} = \bigcap_{j=1,\dots,3N} \mathcal{S}_j$ as we did in Lemma A.4. By Lemma A.4 we then have $(\partial \mathcal{X}_{\mathcal{Q}})_{3j_*} = \partial \mathcal{X}_{\mathcal{S}_{3j_*}}$. Hence, it remains us to describe the stripe $\mathcal{S}_{3j_*}$. For this we recall the precise definition of the extended admissible set $\mathcal{A}_e$, i.e.

$$\mathcal{A}_e = \bigcap_{j \in \mathcal{I} \setminus \mathcal{I}_c} \mathcal{A}^{(j)} \cap \bigcap_{j \in \mathcal{I}_c} \partial \mathcal{A}^{(j)} \cap \mathcal{Q}_e(Z_0) \text{ and } \mathcal{Q}_e = \bigcap_{n=1,\dots,3N} \left\{ z \in \mathbb{R}^{3N} : |(z - Z_0)_n| \leq l_e \right\}.$$

A careful consideration of these sets shows that the $3j_*$-th stripe has the form $\mathcal{S}_{3j_*} = \mathcal{A}^{(j_*)} \cap \left\{ z \in \mathbb{R}^{3N} : |(z - Z_0)_{3j_*}| \leq l_e \right\}$. Due to $j_* \in \mathcal{I}_* \subset \mathcal{I}_s$ we have $(Z_0)_{3j_*} = 0$ and we precise this formula

$$\mathcal{S}_{3j_*} = \left\{ z \in \mathbb{R}^{3N} : 0 \leq z_{3j_*} \leq l_e \right\}.$$

Hence, for the corresponding constraint forces we find

$$(\partial \mathcal{X}_{\mathcal{Q}})_{3j_*} = \partial \mathcal{X}_{\mathcal{S}_{3j_*}} = f\left((z)_{3j_*}\right) - f\left(-(z)_{3j_*} + l_e\right)$$

with the function $f(x) := \begin{cases} 0 & \text{if } x > 0, \\ (-\infty, 0] & \text{if } x = 0. \end{cases}$ Summarizing our results on the $3j_*$-th column of the force balance (5.12) we conclude

$$0 \in \sigma^{(j_*)}(t, z) + f\left((z)_{3j_*}\right) - f\left(-(z)_{3j_*} + l_e\right).$$

(5.13)

But due to the definition of the constraint force f our assumption $\sigma^{(j*)}(t, z) > 0$ implies $(z)_{3j_*} = 0$ or equivalently

$$z \in \partial \mathcal{A}^{(j*)}. \tag{5.14}$$

This proves that the second situation (5.11) implies the first (5.10). We now assume the first situation (5.10) to be satisfied. Our aim is to show that the force balance (5.12), which is equivalent to the second situation, holds. For this we derive the force balance which is equivalent to the first situation and deduce

$$0 \in \mathrm{D}\mathcal{E}(t, z) + \partial_v \left(\Psi_\gamma(t_0, z_0, \cdot - z_0) \right)(z) + \partial \mathcal{X}_{\mathcal{A}_e}(z) + \sum_{j \in \mathcal{I}_*} \partial \mathcal{X}_{\partial \mathcal{A}^{(j)}}(z) \subset \mathbb{R}^{3N^*}. \tag{5.15}$$

We find $\partial \mathcal{X}_{\partial \mathcal{A}^{(j*)}} = \begin{pmatrix} 0 & \cdots & 0 & \mathbb{R} & 0 \cdots & 0 \end{pmatrix}$ as the only difference between (5.12) and (5.15). Hence it is sufficient to show that the $3j_*$-th line of (5.12), i.e. (5.13), is satisfied. In our situation of $z \in \partial \mathcal{A}^{(j*)}$ or $z_{3j_*} = 0$ the inclusion (5.13) simplifies to

$$0 \in \sigma^{(j*)}(t, z) + (-\infty, 0].$$

But this is exactly our additional assumption $\sigma^{(j*)}(t, z) > 0$.   ∎

The following theorem shows the uniform Lipschitz continuity of the approximative solutions. This is the most important and at the same time most challenging result we have to prove. For this reason we want to sketch the central idea behind the result before we give the technical proof in full detail. The difficulty in proving uniform Lipschitz of this result is due to the non-continuity of the dissipation functionals $\Psi^{(j)}$ with respect to z, i.e.

$$\Psi^{(j)}(t, z, v) := \begin{cases} \sigma^{(j)}(t, z)_+ \, \|\mathrm{M}^{(j)}(z^{(j)})v\| & \text{if } z^{(j)} \in \partial \mathcal{S}, \\ 0 & \text{if } z^{(j)} \in \mathrm{int}\mathcal{S}. \end{cases}$$

Why does a general discontinuity of the dissipation functional cause the solution to jump? Our problem is quasi-static or more precisely rate-independent. Hence, at each time the solution $z(t)$ is in a static equilibrium in which the elastic and frictional forces equilibrate. If the dissipation functional and thus the frictional forces decrease instantaneously than the solution jumps to a new equilibrium position.

Why does a physical solution not jump? Our specific dissipation functional $\Psi^{(j)}$ behaves discontinuous when the particle $z^{(j)}$ is making or losing contact. Especially losing contact causes the dissipation functional to decrease discontinuously. But a physical solution will not lose contact as long as it is pressed onto the obstacle with positive normal force, i.e. $\sigma^{(j)} > 0$ holds. Hence, if at time t the particle $z^{(j)}(t) \in \partial \mathcal{S}$ loses contact then we have $\sigma^{(j)} = 0$ and by our specific definition $\Psi^{(j)}(t, z(t), \cdot) \equiv 0$. Thus we pass from zero to zero and no discontinuity occurs.

Unfortunately, our approximative solutions do not show exactly this behavior, but if they are good approximations (when we chose $f_\Pi > 0$ small) then they have a very small normal force $\sigma^{(j)} \ll 1$ when they get and lose contact. Thus they do not behave continuous with respect to $\Psi^{(j)}$ but with respect to our new dissipation functional $\Psi_\gamma^{(j)}$.

Theorem 5.10 (Lipschitz continuity for (IP_γ)) *If the Assumptions* (N0)–(N4) *hold then there exists a Lipschitz constant* $\mathrm{c}_{\mathrm{lip}} > 0$ *and* $\mathrm{C}_{\mathit{fine}} > 0$, *independent of the choice of* γ, *such that the following holds: If* $\gamma > 0$ *and the partition* Π *of* $[T_0, T_0 + \Delta_\mathrm{e}]$ *satisfies*

$$f_\Pi \leq \mathrm{C}_{\mathit{fine}}\gamma,$$

then the solution $(z_k)_{k=0,\dots,N_\Pi}$ *of* (IP_γ) *satisfies the uniform Lipschitz estimate*

$$\|z_k - z_{k-1}\| \leq \mathrm{c}_{\mathrm{lip}}|t_k - t_{k-1}| \quad \text{for } k = 1, \dots, N_\Pi. \tag{5.16}$$

Proof: We define the Lipschitz constant $\mathrm{c}_{\mathrm{lip}} > 0$ via

$$\mathrm{c}_{\mathrm{lip}} := \max\left\{ \frac{\mathrm{c}_1 + \mathrm{c}_2}{\alpha_{\mathcal{K},\mathrm{e}} - \mathrm{q}_{\mathcal{K},\mathrm{e}}} \; : \; \mathcal{I}_\mathrm{c} \subset \mathcal{K} \subset \mathcal{I}_\mathrm{c} \cup \mathcal{I}_\mathrm{s} \right\}. \tag{5.17}$$

See Lemma 5.6 and 5.5 for the definition of the constants involved. Note that $\mathrm{c}_{\mathrm{lip}} > 0$ is independent of the precise choice of $\gamma > 0$. The definition of the upper limit $\mathrm{C}_{\mathrm{fine}}$ of the fineness of the partition we find in (5.20) below.

In the first step we introduce an auxiliary incremental problem. We put $z_0 := Z_0$ and define for $k = 1, \dots, N_\Pi$ incrementally first the set $\mathcal{I}_{k-1} \subset \mathcal{I}_\mathrm{s}$ and then the state $z_k \in \mathcal{A}_\mathrm{e}$ via

$$\mathcal{I}_{k-1} := \left\{ j \in \mathcal{I}_\mathrm{s} \; : \; \sigma^{(j)}(t_{k-1}, z_{k-1}) \geq \frac{\gamma}{2} \right\} \tag{5.18}$$

and

$$z_k \in \mathrm{argmin}\left\{ \mathcal{E}(t_k, y) + \Psi_\gamma(t_{k-1}, z_{k-1}, y - z_{k-1}) \; : \; y \in \mathcal{A}_\mathrm{e} \underset{j \in \mathcal{I}_{k-1}}{\cap} \partial\mathcal{A}^{(j)} \right\}, \tag{5.19}$$

Hence, we restrict a single particle $z^{(j)} \in \mathbb{R}^3$, $j \in \mathcal{I}_\mathrm{s}$ to the boundary when it was pressed onto the boundary with a big normal force $\sigma^{(j)} \geq \gamma/2$ in the previous step. Note that if we replace the set $\mathcal{A}_\mathrm{e} \cap_{j \in \mathcal{I}_{k-1}} \partial\mathcal{A}^{(j)}$ by $\mathcal{A}_\mathrm{e}$ then this auxiliary problem coincides with the Incremental Problem (IP_γ).

We justify this procedure as follows. Let us assume that the auxiliary solution $(z_k)_{k=1,\dots,N_\Pi}$ satisfies for $k = 1, \dots, N$ the Lipschitz estimate (5.16), i.e.

$$\|z_k - z_{k-1}\| \leq \mathrm{c}_{\mathrm{lip}}|t_k - t_{k-1}|.$$

We define the constant $C_\sigma := \sup\left\{ \|\partial_t \sigma^{(j)}\|_{L^\infty(\mathcal{C}_e, \mathbb{R})} + \|D\sigma^{(j)}\|_{L^\infty(\mathcal{C}_e, \mathbb{R}^{3N^*})} : j \in \mathcal{I} \right\}$
such that for all $k = 1, \ldots, N_\Pi$ and $j \in \mathcal{I}$ the following estimate hold $|\sigma^{(j)}(t_k, z_k) - \sigma^{(j)}(t_{k-1}, z_{k-1})| \leq C_\sigma(|t_k - t_{k-1}| + \|z_k - z_{k-1}\|) \leq C_\sigma(1 + c_{\text{lip}})f_\Pi$. Let us choose

$$f_\Pi < C_{\text{fine}}\gamma \quad \text{and} \quad C_{\text{fine}} := \frac{1}{2C_\sigma(1 + c_{\text{lip}})}. \tag{5.20}$$

Let us now consider z_k as defined in (5.19) and take $j \in \mathcal{I}_{k-1}$. Because of the estimates $\sigma^{(j)}(t_{k-1}, z_{k-1}) \geq \gamma/2$ and $|\sigma^{(j)}(t_{k-1}, z_{k-1}) - \sigma^{(j)}(t_k, z_k)| < \gamma/2$ we have

$$\sigma^{(j)}(t_k, z_k) > 0 \quad \text{for all } j \in \mathcal{I}_{k-1}. \tag{5.21}$$

Hence, if our solution satisfies the Lipschitz estimate (5.16) then all particles which are restricted to the boundary are still pressed onto the boundary. Due to the previous Lemma 5.9 and the positive normal forces (5.21) we can replace the set $\mathcal{A}_e \cap_{j \in \mathcal{I}_{k-1}} \partial \mathcal{A}^{(j)}$ by $\mathcal{A}_e$. Summarizing we have shown that if the solution $(z_k)_{k=1,\cdots,N_\Pi}$ of the auxiliary problem (5.19) satisfies for all $k = 1, \cdots, N_\Pi$ the Lipschitz estimate (5.16) then it coincides with the solution of the Incremental Problem (IP$_\gamma$).

In the second part of the proof we show the Lipschitz estimate (5.16) for the auxiliary solution. The proof is carried out by induction. The key argument of the induction is that two consecutive solutions z_k and z_{k+1} formally solve minimizing problems with respect to the same dissipation and admissible set. See for this (5.22) and (5.23) or (5.26) and (5.27) below. This will allow us to apply Lemma 3.18 and to derive recursive estimates as usual. To keep notation short we introduce the dissipation functional

$$\Psi_{\mathcal{I}_c}(t, z, v) := \sum_{j \in \mathcal{I}_c} \Psi^{(j)}(t, z, v).$$

With this notation we have $\Psi_\gamma(T_0, Z_0, \cdot) = \Psi_{\mathcal{I}_c}(T_0, Z_0, \cdot) + \sum_{j \in \mathcal{I}_s} \Psi_\gamma^{(j)}(T_0, Z_0, \cdot)$.
We start our induction and consider the initial values $t_0 = T_0$ and $z_0 = Z_0$. In Lemma 5.5 we have shown that this values satisfy the initial Assumption (3.11) which is equivalent to

$$z_0 \in \operatorname{argmin}\left\{\mathcal{E}(t_0, z) + \Psi_\gamma(t_0, z_0, z - z_0) : z \in \mathcal{A}_e\right\}.$$

Further for all $j \in \mathcal{I}_s$ we have $\sigma^{(j)}(T_0, Z_0) = 0$ or $\Psi_\gamma^{(j)}(t_0, z_0, \cdot) \equiv 0$. Hence, we formally replace Ψ_γ by $\Psi_{\mathcal{I}_c}$, see (5.22) below. The state z_1 is defined via

$$z_1 \in \operatorname{argmin}\left\{\mathcal{E}(t_1, z) + \Psi_{\mathcal{I}_c}(t_0, z_0, z - z_0) + \sum_{j \in \mathcal{I}_0} \Psi_\gamma^{(j)}(t_0, z_0, z^{(j)} - z_0^{(j)}) : \mathcal{A}_e \cap_{j \in \mathcal{I}_0} \partial \mathcal{A}^{(j)}\right\}.$$

But because of the definition of $\mathcal{I}_c = \left\{j \in \mathcal{I} : \sigma^{(j)}(t_0, z_0) > 0\right\}$ and the definition of $\mathcal{I}_0 = \left\{j \in \mathcal{I}_s : \sigma^{(j)}(t_0, z_0) \geq \gamma/2\right\}$ we have $\mathcal{I}_0 = \emptyset$. Hence we find in the definition of z_0 and z_1 the same minimizing set $\mathcal{A}_e$ and the same dissipation functional $\Psi_{\mathcal{I}_c}$, i.e.

$$z_0 \in \operatorname{argmin}\left\{\mathcal{E}(t_0, z) + \Psi_{\mathcal{I}_c}(t_0, z_0, z - z_0) : z \in \mathcal{A}_e\right\} \quad \text{and} \qquad (5.22)$$

$$z_1 \in \operatorname{argmin}\left\{\mathcal{E}(t_1, z) + \Psi_{\mathcal{I}_c}(t_0, z_0, z - z_0) : z \in \mathcal{A}_e\right\}. \qquad (5.23)$$

We are now formally in the setting of Lemma 3.18 of the basic strategy. Note that $\Psi_{\mathcal{I}_c}$ is Lipschitz continuous with respect to $z \in \mathcal{A}_e \subset \mathcal{A} \cap_{j \in \mathcal{I}_c} \partial \mathcal{A}^{(j)}$ with the Lipschitz constant $q_{\mathcal{I}_c,e} > 0$ and $\mathcal{E}(t_0, \cdot)$ is $\alpha_{\mathcal{I}_c,e}$-uniformly convex on $\mathcal{A}_e$. See Lemma 5.6 for the definition of the constants $q_{\mathcal{I}_c,e}, \alpha_{\mathcal{I}_c,e} > 0$. The Lipschitz continuity of $\Psi_{\mathcal{I}_c}$ is the last missing Assumption (3.9) of Lemma 3.18. We derived the Assumptions (3.1)–(3.8) in Lemma 5.5. Applying Lemma 3.18 we find

$$\|z_1 - z_0\| \leq \frac{c_1 + c_2}{\alpha_{\mathcal{I}_c,e}}|t_1 - t_0| \leq \frac{c_1 + c_2}{\alpha_{\mathcal{I}_c,e} - q_{\mathcal{I}_c,e}}|t_1 - t_0| \leq c_{\mathrm{lip}}|t_1 - t_0|.$$

This completes the proof of the induction start.

For the induction step we assume for given $k \in \{1, \cdots, N_{\mathrm{II}}-1\}$ that the Lipschitz estimate (5.16) holds and we will prove

$$\|z_{k+1} - z_k\| \leq c_{\mathrm{lip}}|t_{k+1} - t_k|. \qquad (5.24)$$

Our aim is to show that z_k and z_{k+1} are defined with respect to the same dissipation $\Psi_{\mathcal{I}_c} + \sum_{j \in \mathcal{I}_k} \Psi_\gamma^{(j)}$ and the same set $\mathcal{A}_e \cap_{j \in \mathcal{I}_k} \partial \mathcal{A}^{(j)}$, see (5.26) and (5.27) below. Afterwards we apply again Lemma 3.18 to establish the Lipschitz estimate (5.24). By the definition of the index sets $\mathcal{I}_k, \mathcal{I}_{k-1} \subset \mathcal{I}$, see (5.18), we find

$$z_k \in \operatorname*{argmin}_{\substack{\mathcal{A}_e \cap \partial \mathcal{A}^{(j)} \\ j \in \mathcal{I}_{k-1}}} \left\{\mathcal{E}(t_k, z) + \Psi_{\mathcal{I}_c}(t_{k-1}, z_{k-1}, z - z_{k-1}) + \sum_{j \in \mathcal{I}_{k-1}} \Psi_\gamma^{(j)}(t_{k-1}, z_{k-1}, z - z_{k-1})\right\},$$

$$z_{k+1} \in \operatorname*{argmin}_{\substack{\mathcal{A}_e \cap \partial \mathcal{A}^{(j)} \\ j \in \mathcal{I}_k}} \left\{\mathcal{E}(t_{k+1}, z) + \Psi_{\mathcal{I}_c}(t_k, z_k, z - z_k) + \sum_{j \in \mathcal{I}_k} \Psi_\gamma^{(j)}(t_k, z_k, z - z_k)\right\}.$$

Our aim is to replace the index set $\mathcal{I}_{k-1}$ by $\mathcal{I}_k$ in the definition of z_k. We find for all index $j \in (\mathcal{I}_{k-1} \cup \mathcal{I}_k) \setminus (\mathcal{I}_{k-1} \cap \mathcal{I}_k)$ that the normal force $\sigma^{(j)}$ crosses the level $\gamma/2$, i.e. $\left(\sigma^{(j)}(t_{k-1}, z_{k-1}) - \gamma/2\right)\left(\sigma^{(j)}(t_k, z_k) - \gamma/2\right) \leq 0$. Note that the normal force is formulated with respect to the the states z_{k-1}, z_k for which the Lipschitz estimate (5.16) already holds. As in the first part of the proof we find for their difference $|\sigma^{(j)}(t_k, z_k) - \sigma^{(j)}(t_{k-1}, z_{k-1})| < \gamma/2$ for $f_{\mathrm{II}} < C_{\mathrm{fine}}\gamma$ with $C_{\mathrm{fine}} > 0$ as defined in (5.20). Together with the crossing of the level this proves

$$0 < \sigma^{(j)}(t_k, z_k), \sigma^{(j)}(t_{k-1}, z_{k-1}) < \gamma \qquad (5.25)$$

for all $j \in (\mathcal{I}_{k-1} \cup \mathcal{I}_k) \setminus (\mathcal{I}_{k-1} \cap \mathcal{I}_k)$. Due to the upper bound the normal forces are small, and we make use of our decreased normal force $(\sigma^{(j)} - \gamma)_+$ in the definition of the dissipation functionals $\Psi_\gamma^{(j)}$ to find

$$\Psi_\gamma^{(j)}(t_k, z_k, \cdot) \equiv \Psi_\gamma^{(j)}(t_{k-1}, z_{k-1}, \cdot) \equiv 0.$$

Thus, we have $\sum_{j \in \mathcal{I}_{k-1}} \Psi_\gamma^{(j)} = \sum_{j \in (\mathcal{I}_{k-1} \cap \mathcal{I}_k)} \Psi_\gamma^{(j)} = \sum_{j \in \mathcal{I}_k} \Psi_\gamma^{(j)}$ and we can formally replace the term $\sum_{j \in \mathcal{I}_{k-1}} \Psi_\gamma^{(j)}$ by $\sum_{j \in \mathcal{I}_k} \Psi_\gamma^{(j)}$ in the definition of z_k later on. Thus we can replace the index set $\mathcal{I}_{k-1}$ by $\mathcal{I}_k$ for the dissipation that occurs in the definition of z_k. Next we want to do the same for the admissible set that occurs in the definition of z_k, see (5.26). We achieve this by applying Lemma 5.9. For this we recall the definition of $\mathcal{I}_{k-1} = \{j \in \mathcal{I}_s : \sigma^{(j)}(t_{k-1}, z_{k-1}) > \gamma/2\}$ and hence we can replace $\Psi_{\mathcal{I}_c} + \sum_{j \in \mathcal{I}_{k-1}} \Psi_\gamma^{(j)}$ by $\Psi_\gamma = \Psi_{\mathcal{I}_c} + \sum_{j \in \mathcal{I}_s} \Psi_\gamma^{(j)}$, see (5.10) and (5.11). Putting $\mathcal{I}_* = (\mathcal{I}_{k-1} \cup \mathcal{I}_k) \subset \mathcal{I}_s$ we find $\sigma^{(j)}(t_k, z_k) > 0$ for all $j \in \mathcal{I}_*$ and we are formally in the setting of Lemma 5.9. We use the equivalence of Lemma 5.9 in both directions. First we pass from (5.11) to (5.10) to increase the index set $\mathcal{I}_{k-1}$ to $\mathcal{I}_* = \mathcal{I}_{k-1} \cup \mathcal{I}_k$ and then use the other direction to decrease the index set again to $\mathcal{I}_k$. Thus we have shown that we can replace the index set $\mathcal{I}_{k-1}$ by $\mathcal{I}_k$ in the definition of z_k and we formally find

$$z_k \in \underset{\mathcal{A}_e \underset{j \in \mathcal{I}_k}{\cap} \partial \mathcal{A}^{(j)}}{\operatorname{argmin}} \left\{ \mathcal{E}(t_k, z) + \Psi_{\mathcal{I}_c}(t_{k-1}, z_{k-1}, z - z_{k-1}) + \sum_{j \in \mathcal{I}_k} \Psi_\gamma^{(j)}(t_{k-1}, z_{k-1}, z - z_{k-1}) \right\},$$
(5.26)

$$z_{k+1} \in \underset{\mathcal{A}_e \underset{j \in \mathcal{I}_k}{\cap} \partial \mathcal{A}^{(j)}}{\operatorname{argmin}} \left\{ \mathcal{E}(t_{k+1}, z) + \Psi_{\mathcal{I}_c}(t_k, z_k, z - z_k) + \sum_{j \in \mathcal{I}_k} \Psi_\gamma^{(j)}(t_k, z_k, z - z_k) \right\}.$$
(5.27)

We finally have shown that z_k and z_{k+1} are defined with respect to the same dissipation $\Psi_{\mathcal{I}_c} + \sum_{j \in \mathcal{I}_k} \Psi_\gamma^{(j)}$ and the same admissible set $\mathcal{A}_e \cap_{j \in \mathcal{I}_k} \partial \mathcal{A}^{(j)}$. The combination is such that the dissipation is Lipschitz continuous with respect to the second variable z if z is restricted to the above admissible set. For $\mathcal{K} := \mathcal{I}_c \cup \mathcal{I}_k$ we find the Lipschitz constant $q_{\mathcal{K},e}$. See Lemma 5.6 for the precise definition of the constants $q_{\mathcal{K},e}, \alpha_{\mathcal{K},e} > 0$. The Lipschitz continuity allows us to apply Lemma 3.18. Due to the equality $|t_{k+1} - t_k| = |t_k - t_{k-1}|$ we find

$$\|z_{k+1} - z_k\| \leq \frac{(c_1 + c_2)}{\alpha_{\mathcal{K},e}} |t_{k+1} - t_k| + \frac{q_{\mathcal{K},e}}{\alpha_{\mathcal{K},e}} \|z_k - z_{k-1}\|$$

By estimate (5.16) we deduce $\|z_{k+1} - z_k\| \leq \left(\frac{(c_1 + c_2)}{\alpha_{\mathcal{K},e}} + \frac{q_{\mathcal{K},e}}{\alpha_{\mathcal{K},e}} c_{\mathrm{lip}} \right) |t_{k+1} - t_k|$. It remains to prove $\left(\frac{(c_1 + c_2)}{\alpha_{\mathcal{K},e}} + \frac{q_{\mathcal{K},e}}{\alpha_{\mathcal{K},e}} c_{\mathrm{lip}} \right) \leq c_{\mathrm{lip}}$. This is equivalent to $(c_1 + c_2) \leq (\alpha_{\mathcal{K},e} - q_{\mathcal{K},e}) c_{\mathrm{lip}}$.

Due to the extended convexity assumption $\alpha_{\mathcal{K},e} > q_{\mathcal{K},e}$ the inequality does not change if we divide by $(\alpha_{\mathcal{K},e} - q_{\mathcal{K},e}) > 0$. By the definition of c_{lip} in (5.17) we find

$$\|z_{k+1} - z_k\| \le c_{lip}|t_{k+1} - t_k|.$$

We concluded the induction step. Hence, the solution $(z_k)_{k=1,\ldots,N_\Pi}$ of the Auxiliary Problem (5.19) satisfies the Lipschitz estimate (5.16). In the first part of the proof we have shown that this implies that the auxiliary solution coincides with the solution of (IP_γ). ∎

For $j \in I_s$ we introduce the closed domain

$$\mathcal{D}^{(j)} := \left\{ (t,z) \in [T_0, T_0 + \Delta_e] \times \mathcal{A}_e \ : \ \sigma^{(j)}(t,z) \le 0 \text{ or } z^{(j)} \in \partial \mathcal{S} \right\}.$$

The domain $\mathcal{D}$ it then defined via

$$\mathcal{D} := ([T_0, T_0 + \Delta_e] \times \mathcal{A}_e) \cap \left(\bigcap_{j \in I_s} \mathcal{D}^{(j)} \right). \tag{5.28}$$

Lemma 5.11 (Continuity of $\Psi_\gamma^{(j)}$) *Let the Regularity Assumptions* (N1), (N2) *hold for $\mathcal{E}$ and* $\mathrm{M}^{(j)}$. *Then we have*

$$\Psi_\gamma^{(j)} \in C(\mathcal{D}^{(j)} \times \mathbb{R}^{3N}, [0, \infty)) \quad \text{for all } j = 1, \ldots, N \text{ and } \gamma \ge 0.$$

Proof: We recall the definition

$$\Psi_\gamma^{(j)}(t,z,v) = \begin{cases} \left(\sigma^{(j)}(t,z) - \gamma \right)_+ \|\mathrm{M}^{(j)}(z^{(j)})v\| & \text{if } z^{(j)} \in \partial \mathcal{S}, \\ 0 & \text{if } z^{(j)} \in \text{int} \mathcal{S}. \end{cases}$$

The function $\Psi_\gamma^{(j)}$ is upper semi-continuous and positive. Thus it is sufficient to consider a sequence $(t_n, z_n, v_n)_{n \in \mathbb{N}} \subset \mathcal{D}^{(j)} \times \mathbb{R}^3$ with a limit $(t,z,v) \in \mathcal{D}^{(j)} \times \mathbb{R}^3$ that satisfies $\Psi_\gamma^{(j)}(t,z,v) > 0$. This implies $z^{(j)} \in \partial \mathcal{S}$ and $\sigma^{(j)}(t,z) > \gamma$. Since $\sigma^{(j)}$ is continuous there exists $n_0 \in \mathbb{N}$ such that $\sigma^{(j)}(t_n, z_n) > 0$ for all $n \ge n_0$. By the definition of $\mathcal{D}^{(j)}$ this implies $z_n^{(j)} \in \partial \mathcal{S}$ for all $n \ge n_0$. Thus for $n \ge n_0$ we find $\Psi_\gamma^{(j)}(t_n, z_n, v_n)$ and $\Psi_\gamma^{(j)}(t,z,v)$ being both defined by the continuous function $\left(\sigma^{(j)}(t,z) - \gamma \right)_+ \|\mathrm{M}^{(j)}(z^{(j)})v\|$. ∎

Theorem 5.12 (Existence for Problem 5.4) *Let us assume* (N0)–(N4). *Then there exists a constant $c_{lip} > 0$ independent of γ and a solution $z_\gamma \in W^{1,\infty}([T_0, T_0 + \Delta_e], \mathcal{A}_e)$ of Problem 5.4 satisfying for all $\tau_1, \tau_2 \in [T_0, T_0 + \Delta_e]$*

$$\|z_\gamma(\tau_2) - z_\gamma(\tau_1)\| \le c_{lip}|\tau_2 - \tau_1| \quad \text{and} \quad (t, z_\gamma(t)) \in \mathcal{D} \tag{5.29}$$

with $\mathcal{D}$ as defined in (5.28).

Proof: We prove existence with the help of our generic Existence Theorem 3.26 of the basic strategy. We have to verify its Assumptions (3.2), (3.4)–(3.6) and (3.31)–(3.34). The typical properties (3.2) and (3.4)–(3.6) of $\mathcal{E}$ and Ψ_γ follow directly from our Assumptions (N0)–(N4) as we have shown in Lemma 5.5. Next a sequence of partitions with $f_{\Pi^{(n)}} \to 0$ was assumed to be given in Theorem 3.26. We have proven in Theorem 5.8 the existence and uniqueness of the corresponding sequence of discrete solutions $(z_k^{(n)})_{k=1,\dots,N_{\Pi^{(n)}}}$ of the incremental problem (IP$_\gamma$). This is (3.31). In Theorem 5.10 we verified (3.32), i.e. the uniform Lipschitz continuity of the approximative solutions $\big(z_\gamma{}^{(n)}\big)_{n \in \mathbb{N}} \subset \mathrm{W}^{1,\infty}([T_0, T_0 + \Delta_\mathrm{e}], \mathcal{A}_\mathrm{e})$. Note that the Lipschitz constant c_lip defined in (5.17) is independent of γ. The existence of a convergent subsequence, i.e. (3.33), is a consequence of Theorem 3.15.

It remains us to check the last Assumption (3.34), i.e. the existence of a domain $\mathcal{D} \subset [T_0, T_0 + \Delta_\mathrm{e}] \times \mathcal{A}_\mathrm{e}$ such that $(t_k, z_k) \in \mathcal{D}$ holds for all solutions of (IP$_\gamma$) and such that $\Psi_\gamma = \sum_{j \in \mathrm{I_c}} \Psi^{(j)} + \sum_{j \in \mathrm{I_s}} \Psi_\gamma^{(j)} \in \mathrm{C}(\mathcal{D} \times \mathbb{R}^n, \mathbb{R})$. We choose $\mathcal{D}$ as in (5.28). We first prove $(t_k, z_k) \in \mathcal{D}$ for a solution $(z_k)_{k=1,\dots,N_\Pi}$ of (IP$_\gamma$). It is sufficient to prove for $j \in \mathrm{I_s}$ that $z_k^{(j)} \notin \partial\mathcal{S}$ implies $\sigma^{(j)}(t, z) \leq 0$. In Figure 5.3 we present the

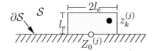

Figure 5.3: Minimizing region for $z_k^{(j)}$ and for $j \in \mathrm{I_s}$.

minimizing region for $z_k^{(j)} \in \mathcal{S}$ with $j \in \mathrm{I_s}$. See also Figure 5.2. Thus $z_k^{(j)} \notin \partial\mathcal{S}$ implies the existence of $\lambda_0 > 0$ such that $y := z_k + \lambda(-\mathrm{e}_{3j}) \in \mathcal{A}_\mathrm{e}$ for all $0 \leq \lambda \leq \lambda_0$ and since z_k is a minimizer on $\mathcal{A}_\mathrm{e}$ we find

$$\mathcal{E}\big(t_k, z_k\big) + \tilde{\Psi}_\gamma\big(z_k - z_{k-1}\big) \leq \mathcal{E}\big(t_k, z_k + \lambda(-\mathrm{e}_{3j})\big) + \tilde{\Psi}_\gamma\big(z_k - z_{k-1} + \lambda(-\mathrm{e}_{3j})\big)$$

for all $0 \leq \lambda \leq \lambda_0$ and for the function $\tilde{\Psi}_\gamma(v) := \Psi_\gamma(t_{k-1}, z_{k-1}, v)$. Due to $\mathrm{M}^{(j)}(z)\mathrm{e}_{3j}^{(j)} = 0$, see (N2), the dissipation functionals coincide and we have, after division with λ,

$$0 \leq \frac{1}{\lambda}\Big(\mathcal{E}(t_k, z_k + \lambda(-\mathrm{e}_{3j})) - \mathcal{E}(t_k, z_k + \lambda(-\mathrm{e}_{3j}))\Big).$$

Passing to the limit $\lambda \to 0$ we find $0 \leq \mathrm{D}\mathcal{E}(t_k, z_k)(-\mathrm{e}_{3j}) = -\sigma^{(j)}(t_k, z_k)$. This proves $(t_k, z_k) \in \mathcal{D}$.

We still have to check the second property of Assumption (3.34), i.e. $\Psi_\gamma = \sum_{j \in \mathrm{I_c}} \Psi^{(j)} + \sum_{j \in \mathrm{I_s}} \Psi_\gamma^{(j)} \in \mathrm{C}(\mathcal{D} \times \mathbb{R}^{3N}, \mathbb{R})$. Since $z \in \mathcal{A}_\mathrm{e}$ implies $z^{(j)} \in \partial\mathcal{S}$ for $j \in \mathrm{I_c}$ we find $\Psi^{(j)}$ to be continuous on the bigger domain $[T_0, T_0 + \Delta_\mathrm{e}] \times \mathcal{A}_\mathrm{e} \times \mathbb{R}^{3N}$.

for all $j \in I_c$. And due to $\mathcal{D} \subset \mathcal{D}^{(j)}$ it is thus sufficient to have $\Psi_\gamma^{(j)} \in C\big(\mathcal{D}^{(j)} \times \mathbb{R}^3, \mathbb{R}\big)$ for $j \in I_s$, see Lemma 5.11. This proves the continuity of Ψ_γ on $\mathcal{D} \times \mathbb{R}^{3N}$ and shows that the Assumption (3.34) holds.

We now verified all assumptions of the Existence Theorem 3.26 and establish our result.   ∎

5.5 Solving the problem

In this section we solve the N particle Problem 5.1. We construct a solution considering a sequence of simplified solutions z_{γ_n} whose parameters converge to zero, i.e $\gamma_n \to 0$ for $n \to 0$.

Theorem 5.13 *We assume* (N0)–(N4). *Then there exists a time span* $\Delta \in (0, T - T_0]$ *and a solution* $z \in W^{1,\infty}([T_0, T_0{+}\Delta], \mathcal{A})$ *of Problem 5.1, i.e. for almost all* $t \in [T_0, T_0{+}\Delta]$ *we have*

$$0 \in D\mathcal{E}(t, z(t)) + \partial_v \Psi(t, z(t), \dot{z}(t)) + \mathcal{N}_\mathcal{A}(z(t)) \quad \subset \mathbb{R}^{3N^*}. \tag{DI}$$

Proof: We consider a sequence of parameters $(\gamma_n)_{n \in \mathbb{N}} \subset (0, \infty)$ with $\gamma_n \to 0$ for $n \to \infty$. We denote by $(z_{\gamma_n})_{n \in \mathbb{N}} \subset W^{1,\infty}([T_0, T_0 + \Delta_e], \mathcal{A}_e)$ the sequence of solutions of the corresponding simplified Problems 5.4. The Existence Theorem 5.12 shows the existence of the solutions and provides the uniform Lipschitz estimate (5.29), i.e.

$$\|z_{\gamma_n}(\tau_2) - z_{\gamma_n}(\tau_1)\| \le c_{\text{lip}}|\tau_2 - \tau_1| \quad \text{and} \quad (\tau, z_{\gamma_n}(\tau)) \in \mathcal{D}$$

for all $\tau_1, \tau_2, \tau \in [T_0, T_0 + \Delta_e]$ and $n \in \mathbb{N}$. Since $c_{\text{lip}} > 0$ is independent of $n \in \mathbb{N}$ and all solutions satisfy the same initial condition $z_{\gamma_n}(T_0) = Z_0$ there exists, due to Theorem 3.15, a convergent subsequence, which we still denote by $(z_{\gamma_n})_{n \in \mathbb{N}} \subset W^{1,\infty}([T_0, T_0 + \Delta_e], \mathcal{A}_e)$ and a limit function $z \in W^{1,\infty}([T_0, T_0 + \Delta_e], \mathcal{A}_e)$ such that

$$\|z_{\gamma_n} - z\|_{L^\infty([T_0, T_0+\Delta_e], \mathcal{A}_e)} \to 0 \quad \text{for } n \to \infty. \tag{5.30}$$

In the following we prove that z is our desired solution. But let us present a further convergence result first. There exists $c_{\text{te}} > 0$ such that for all $j \in \mathcal{I}, z \in \mathcal{A}_e$ and $v \in \mathbb{R}^{3N}$ we have

$$\left|\Psi_\gamma^{(j)}(t, z, v) - \Psi^{(j)}(t, z, v)\right| \le c_{\text{te}} \|v\| \gamma. \tag{5.31}$$

By definition this difference equals $\left|\big(\sigma^{(j)}(t, z) - \gamma\big)_+ - \sigma^{(j)}(t, z)_+\right| \|M^{(j)}(z^{(j)})v\|$. Finally we define $c_{\text{te}} := \sup \big\{\|M^{(j)}(z^{(j)})\| : j \in \mathcal{I}, z \in \mathcal{A}_e\big\}$ such that the estimate (5.31) follows directly.

We next show that z satisfies an (S)&(E)-formulation. For all $n \in \mathbb{N}$ and $t \in [T_0, T_0 + \Delta_e]$ we have

$$\mathcal{E}(t, z_{\gamma_n}(t)) \leq \mathcal{E}(t, y) + \Psi_{\gamma_n}(t, z_{\gamma_n}(t), y - z_{\gamma_n}(t)) \quad \text{for all } y \in \mathcal{A}_e \quad \text{and} \qquad (\mathrm{S}_n)$$

$$\mathcal{E}(t, z_{\gamma_n}(t)) + \int_{T_0}^{t} \Psi_{\gamma_n}(s, z_{\gamma_n}(s), \dot{z}_{\gamma_n}(s)) \, \mathrm{d}s = \mathcal{E}(T_0, Z_0)) + \int_{T_0}^{t} \partial_s \mathcal{E}(s, z_{\gamma_n}(s)) \, \mathrm{d}s. \qquad (\mathrm{E}_n)$$

Passing to the limit we see due to (5.30) and (N1) that we can replace z_{γ_n} by z in all energy terms. Concerning the dissipational terms we replace Ψ_{γ_n} by $\Psi_{c,s} := \sum_{j \in \mathcal{I}_c \cup \mathcal{I}_s} \Psi^{(j)}$. Note that we have $\Psi_\gamma - \Psi_{c,s} = \sum_{j \in \mathcal{I}_s} \Psi_\gamma^{(j)} - \Psi^{(j)}$. The dissipation in (S_n) is estimated as follows (we drop any dependence on t to keep the formula short), for $j \in \mathcal{I}_s$ we have

$$|\Psi_{\gamma_n}^{(j)}(z_{\gamma_n}, y - z_{\gamma_n}) - \Psi^{(j)}(z, y - z)| \leq |\Psi_{\gamma_n}^{(j)}(z_{\gamma_n}, y - z_{\gamma_n}) - \Psi^{(j)}(z_{\gamma_n}, y - z_{\gamma_n})|$$
$$+ |\Psi^{(j)}(z_{\gamma_n}, y - z_{\gamma_n}) - \Psi^{(j)}(z, y - z)|.$$

The first difference converges due to (5.31). For the second difference we exploit $(t, z_{\gamma_n}(t)) \in \mathcal{D}$ and $\Psi^{(j)} \in C(\mathcal{D} \times \mathbb{R}^{3N}, \mathbb{R})$, see Lemma 5.11. Hence the uniform convergence (5.30) is sufficient to pass to the limit and to find for all $t \in [T_0, T_0 + \Delta]$ the stability condition

$$\mathcal{E}(t, z(t)) \leq \mathcal{E}(t, y) + \Psi_{c,s}(t, z(t), y - z(t)) \quad \text{for all } y \in \mathcal{A}_e. \qquad (\mathrm{S}_{c,s})$$

Applying Lemma 3.24 we deduce for all $t \in [T_0, T_0 + \Delta]$ the lower energy estimate

$$\mathcal{E}(t, z(t)) + \int_{T_0}^{t} \Psi_{c,s}(s, z(s), \dot{z}(s)) \, \mathrm{d}s \geq \mathcal{E}(T_0, Z_0)) + \int_{T_0}^{t} \partial_s \mathcal{E}(s, z(s)) \, \mathrm{d}s.$$

It remains us to derive the opposite estimate and to treat the integral in (E_n). We replace the integrand $\Psi_{\gamma_n}(s, z_{\gamma_n}(s), \dot{z}_{\gamma_n}(s))$ by $\Psi_{c,s}(s, z(s), \dot{z}_{\gamma_n}(s))$ doing the same estimates as above. Taking the lim inf in (E_n), see also Lemma 3.24, we find for the dissipational integral

$$\int_{T_0}^{t} \Psi_{c,s}(s, z(s), \dot{z}(s)) \, \mathrm{d}s \leq \liminf_{n \to \infty} \int_{T_0}^{t} \Psi_{c,s}(s, z(s), \dot{z}_{\gamma_n}(s)) \, \mathrm{d}s.$$

We thus established the upper energy estimate and together with the above lower estimate this proves

$$\mathcal{E}(t, z(t)) + \int_{T_0}^{t} \Psi_{c,s}(s, z(s), \dot{z}(s)) \, \mathrm{d}s = \mathcal{E}(T_0, Z_0)) + \int_{T_0}^{t} \partial_s \mathcal{E}(s, z(s)) \, \mathrm{d}s. \qquad (\mathrm{E}_{c,s})$$

All in all, we have shown, that the function z satisfies an $(\text{S}_{\text{c,s}})\&(\text{E}_{\text{c,s}})$-formulation. In Theorem 2.9 we have shown that due to the convexity of $\mathcal{E}(t,\cdot)$ on $\mathcal{A}_{\text{e}}$ for all $t \in [T_0, T_0 + \Delta_{\text{e}}]$ the $(\text{S}_{\text{c,s}})\&(\text{E}_{\text{c,s}})$-formulation is equivalent to

$$0 \in \text{D}\mathcal{E}(t, z(t)) + \partial_v \Psi_{\text{c,s}}(t, z(t), \dot{z}(t)) + \mathcal{N}_{\mathcal{A}_{\text{e}}}(z(t)) \quad \subset \mathbb{R}^{3N^*}.$$

We finally shorten the time span Δ_{e} such that those particles which are pressed onto the obstacle initially remain in contact and those particles which are out of contact at time T_0 remain friction-free. We first choose $\Delta \leq \min\{l_{\text{e}}/(2c_{\text{lip}}), \Delta_{\text{e}}]$ such that $z(t) \in \text{int}\mathcal{Q}_{l_{\text{e}}}(Z_0)$ holds for all $t \in [T_0, T_0 + \Delta]$. Hence, the solution does not touch the box $\mathcal{Q}_{l_{\text{e}}}$. This allows us to replace formally the constraint forces $\mathcal{N}_{\mathcal{A}_{\text{e}}} = \mathcal{N}_{\mathcal{A}} + \sum_{j \in \mathcal{I}_{\text{c}}} \mathcal{N}_{\partial \mathcal{A}^{(j)}} + \mathcal{N}_{\mathcal{Q}_{l_{\text{e}}}(Z_0)}$ by $\mathcal{N}_{\mathcal{A}} + \sum_{j \in \mathcal{I}_{\text{c}}} \mathcal{N}_{\partial \mathcal{A}^{(j)}}$. Further, for $j \in \mathcal{I}_{\text{c}}$ we have by definition $\sigma^{(j)}(T_0, Z_0) > 0$ and we further shorten $\Delta \in (0, \Delta_{\text{e}}]$ such that $\sigma^{(j)}(t, z(t)) \geq 0$ holds for all $j \in \mathcal{I}_{\text{c}}$ and $t \in [T_0, T_0 + \Delta]$. Note that the corresponding column in the above force balance law is the $3j$-th column and it reads

$$0 \in \sigma^{(j)}(t, z(t)) + \{0\} + (-\infty, 0] + \mathbb{R}^*.$$

The zero corresponds to $\partial_v \Psi_{\text{c,s}}$, the half axes to $\mathcal{N}_{\mathcal{A}}$ and the whole axes $\mathbb{R}^*$ is the only non-zero contribution of the constraint force $\mathcal{N}_{\mathcal{A}^{(j)}}$. Hence, due to $\sigma^{(j)} > 0$ we see that we can skip the contribution of $\mathcal{N}_{\mathcal{A}^{(j)}}$ and the system still holds. This proves

$$0 \in \text{D}\mathcal{E}(t, z(t)) + \partial_v \Psi_{\text{c,s}}(t, z(t), \dot{z}(t)) + \mathcal{N}_{\mathcal{A}}(z(t)) \quad \subset \mathbb{R}^{3N^*}$$

for almost all times $t \in [T_0, T_0 + \Delta]$. For $j \in \mathcal{I}_{\text{f}}$ we have $Z_0^{(j)} \in \text{int}\mathcal{A}^{(j)}$ and since z is Lipschitz continuous there exists a time span $\Delta \in (0, \Delta_{\text{e}}]$ such that $z^{(j)}(t) \in \text{int}\mathcal{A}^{(j)}$ holds for all $j \in \mathcal{I}_{\text{f}}$ and all $t \in [T_0, T_0 + \Delta)$. By the definition of $\Psi^{(j)}$ this implies $\Psi^{(j)}(t, z(t), \cdot) \equiv 0$ for $j \in \mathcal{I}_{\text{f}}$ and $t \in [T_0, T_0 + \Delta]$. Hence we can replace $\Psi_{\text{c,s}}$ by Ψ in the above inclusion. And we established our result. ∎

Chapter 6

Curved boundary - the full problem

Figure 6.1: Non-linear geometry: non-convex set $\mathcal{A}$

6.1 Overview

In this chapter we consider the full problem, i.e. varying frictional forces, making and losing contact and curved boundaries. The main results are Theorem 6.5 and 6.19. Up to now we considered admissible sets $\bar{\mathcal{S}} = \{z \in \mathbb{R}^3 : z_3 \geq 0\}$ with a flat obstacle boundary $\partial \bar{\mathcal{S}} = \{z \in \mathbb{R}^3 : z_3 = 0\}$. Here we introduced the bar ' $^-$ ' to indicate the flat case. In this section we want to extend our existence results to the situation of a curved boundary which can be described by the graph of a shape function $\varphi_0 \in C^2(\mathbb{R}^2, \mathbb{R})$. For example one could imagine admissible sets of the form $\mathcal{S} = \{z \in \mathbb{R}^3 : z_3 \geq \varphi_0(z_1, z_2)\}$.

Our main task is to examine the influence of the new data (the shape function φ_0) on our assumptions for existence. We will achieve this with the help of a flattening transformation $\Phi \in C^2(\mathbb{R}^3, \mathbb{R}^3)$ which is locally bijective and maps the flat boundary $\partial \bar{\mathcal{S}} = \{z \in \mathbb{R}^3 : z_3 = 0\}$ on the general curved boundary $\partial \mathcal{S}$. This flattening transformation Φ allows us to reduce the situation of the curved boundary to the situation of a flat boundary and to use our previous results of Chapter 4 and 5 again.

6.2   Modeling a curved boundary

Definition 6.1 (C^k-boundary, normal vector and tangential space)
Let the set $S \subset \mathbb{R}^3$ satisfy $S = \mathrm{clos}(\mathrm{int}S)$ and $k \in \mathbb{N}$ be given. We say that the boundary ∂S is locally C^k if the following holds. For every $Z_0 \in \partial S$ there exists an open and connected set $\mathcal{U}_0 \subset \mathbb{R}^3$ with $Z_0 \in \mathcal{U}_0$, a radius $r_0 > 0$, an orthonormal matrix $A_0 \in \mathbb{R}^{3 \times 3}$ called rotation *and a shape function $\varphi_0 \in C^k(\mathbb{R}^2, \mathbb{R})$ with*

$$\varphi_0(0) = 0 \quad and$$
$$\mathrm{D}\varphi_0(0) = 0 \in \mathbb{R}^{2^*}$$

such that the local representation $\Gamma_0 \in C^k(\mathrm{int}\mathcal{B}_{r_0}(0), \mathcal{U}_0)$ *defined for $\bar{z} \in \mathrm{int}\mathcal{B}_{r_0}(0) \subset$*

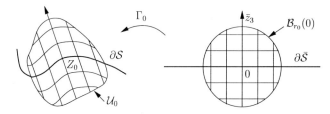

Figure 6.2: The local representation Γ_0

$\mathbb{R}^3$ by

$$\Gamma_0(\bar{z}) := Z_0 + A_0 \begin{pmatrix} \bar{z}_1 \\ \bar{z}_2 \\ \bar{z}_3 + \varphi_0(\bar{z}_1, \bar{z}_2) \end{pmatrix}$$

has the following properties. See also Figure 6.2. Γ_0 is a bijection between the sets

$$\{\bar{z} \in \mathrm{int}\mathcal{B}_{r_0}(0) : \bar{z}_3 > 0\} \quad and \quad \mathrm{int}S \cap \mathcal{U}_0,$$
$$\{\bar{z} \in \mathrm{int}\mathcal{B}_{r_0}(0) : \bar{z}_3 = 0\} \quad and \quad \partial S \cap \mathcal{U}_0 \quad and \ finally$$
$$\{\bar{z} \in \mathrm{int}\mathcal{B}_{r_0}(0) : \bar{z}_3 < 0\} \quad and \quad \mathcal{U}_0/S.$$

For $z \in \partial S \cap \mathcal{U}_0$ the outward normal vector $\nu(z)$ is given by

$$\nu(z) := \frac{1}{\|(\partial_{z_1}\varphi_0, \partial_{z_2}\varphi_0, -1)(\bar{z}_1, \bar{z}_2)\|} A_0 \begin{pmatrix} \partial_{z_1}\varphi_0(\bar{z}_1, \bar{z}_2) \\ \partial_{z_2}\varphi_0(\bar{z}_1, \bar{z}_2) \\ -1 \end{pmatrix} \quad for \ z = \Gamma_0(\bar{z}_1, \bar{z}_2, 0).$$

Remark 6.2 *The local representation Γ_0 is unique up to some orthonormal matrix $B \in \mathbb{R}^{2 \times 2}$. More precisely let us assume that $\tilde{\Gamma}_0(\bar{z}) := Z_0 + \tilde{A}_0 \begin{pmatrix} \bar{z}_1 \\ \bar{z}_2 \\ \bar{z}_3 + \tilde{\varphi}_0(\bar{z}_1, \bar{z}_2) \end{pmatrix}$*

and $\hat{\Gamma}_0(\bar{z}) := Z_0 + \hat{A}_0 \begin{pmatrix} \bar{z}_1 \\ \bar{z}_2 \\ \bar{z}_3 + \hat{\varphi}_0(\bar{z}_1, \bar{z}_2) \end{pmatrix}$ both are local representations for some

given $Z_0 \in \partial \mathcal{S}$, then there exists some orthonormal matrix $B \in \mathbb{R}^{2 \times 2}$ such that we

have $\tilde{\varphi}_0(\tilde{z}_1, \tilde{z}_2) = \hat{\varphi}_0(\hat{z}_1, \hat{z}_2)$ for $B \begin{pmatrix} \tilde{z}_1 \\ \tilde{z}_2 \end{pmatrix} = \begin{pmatrix} \hat{z}_1 \\ \hat{z}_2 \end{pmatrix}$ and $\tilde{A}_0 = \hat{A}_0 \begin{pmatrix} b_{11} & b_{21} & 0 \\ b_{12} & b_{22} & 0 \\ 0 & 0 & 1 \end{pmatrix}$.

The next lemma presents two representations for the tangential plane $\mathcal{T}_{\partial \mathcal{S}}(z)$ and shows implicitly that the definition of the normal vector $\nu(z)$ is independent of the chosen local representation Γ_0, see (6.3).

Lemma 6.3 *We assume the set $\mathcal{S} \subset \mathbb{R}^3$ to have a C^1-boundary, $z \in \partial \mathcal{S}$ and a local representation Γ_0 to be given with $\Gamma_0(\bar{z}) = z$ for some $\bar{z} \in \partial \bar{\mathcal{S}}$. Then we find*

$$\mathcal{T}_{\partial \mathcal{S}}(z) = \mathrm{D}\Gamma_0(\bar{z})\mathcal{T}_{\partial \bar{\mathcal{S}}}(\bar{z}), \tag{6.1}$$

$$= \left\{ v \in \mathbb{R}^3 \ : \ \nu^\top(z)v = 0 \right\} \quad and \tag{6.2}$$

$$\mathcal{N}_{\mathcal{S}}(z) = \left\{ v^* \in \mathbb{R}^{3^*} \ : \ v^* = \lambda \nu^\top(z), \lambda \geq 0 \right\}. \tag{6.3}$$

Proof: The set $\mathcal{T}_{\partial \mathcal{S}}(z)$, which we call the tangential plane, was defined in Section 2.1.4 as a tangential cone $\mathcal{T}_{\partial \mathcal{S}}(z) := \cap_{\varepsilon > 0} \mathcal{T}_{\partial \mathcal{S}}^\varepsilon(z)$ with the help of the sets

$$\mathcal{T}_{\partial \mathcal{S}}^\varepsilon(z) = \mathrm{clos} \left\{ v \in \mathbb{R}^3 \ : \ \exists \lambda > 0 \text{ such that } y = \lambda v + z \in \partial \mathcal{S} \cap \mathcal{B}_\varepsilon(z) \right\}.$$

Due to the simple structure of the flat boundary $\partial \bar{\mathcal{S}}$ we find for $\bar{z} \in \partial \bar{\mathcal{S}}$ the simple relations $\mathcal{T}_{\partial \bar{\mathcal{S}}}^\varepsilon(\bar{z}) = \mathcal{T}_{\partial \bar{\mathcal{S}}}(\bar{z}) = \{v \in \mathbb{R}^3 \ : \ v_3 = 0\}$. Because of $\det(\mathrm{D}\Gamma_0) \equiv 1$ there exist constants $0 < c \leq C$ such that we have for

$$\bar{\mathcal{T}}_{\partial \mathcal{S}}^\varepsilon(z) := \mathrm{clos} \left\{ v \in \mathbb{R}^3 \ : \ \exists \lambda > 0 \text{ s. th. } \lambda v = (\Gamma_0(\bar{y}) - \Gamma_0(\bar{z})) \text{ for } \bar{y} \in \partial \bar{\mathcal{S}} \cap \mathcal{B}_\varepsilon(\bar{z}) \right\}$$

the inclusions $\mathcal{T}_{\partial \mathcal{S}}^{c\varepsilon}(z) \subset \bar{\mathcal{T}}_{\partial \mathcal{S}}^\varepsilon(z) \subset \mathcal{T}_{\partial \mathcal{S}}^{C\varepsilon}(z)$. Hence, we replace $\mathcal{T}_{\partial \mathcal{S}}^\varepsilon(z)$ by $\mathcal{T}_{\partial \mathcal{S}}^\varepsilon(z)$ in the definition of $\mathcal{T}_{\partial \mathcal{S}}(z)$ and thus rewrite

$$\mathcal{T}_{\partial \mathcal{S}}(z) = \underset{\varepsilon > 0}{\cap} \mathrm{clos} \left\{ v \in \mathbb{R}^3 \ : \ \exists \lambda > 0, \lambda v = (\Gamma_0(\bar{y}) - \Gamma_0(\bar{z})) \text{ for } \bar{y} \in \partial \bar{\mathcal{S}} \cap \mathcal{B}_\varepsilon(\bar{z}) \right\}$$

$$= \left\{ v \in \mathbb{R}^3 \ : \ v = \mathrm{D}\Gamma_0(\bar{z})w \text{ for } w \in \mathbb{R}^3, w_3 = 0 \right\} = \mathrm{D}\Gamma_0(\bar{z})\mathcal{T}_{\partial \bar{\mathcal{S}}}(\bar{z}).$$

This proves the first equality (6.1), i.e. $\mathcal{T}_{\partial \mathcal{S}}(z) = \mathrm{D}\Gamma_0(\bar{z})\mathcal{T}_{\partial \bar{\mathcal{S}}}(\bar{z})$. In the same way one can prove the equality $\mathcal{T}_{\mathcal{S}}(z) = \mathrm{D}\Gamma_0(\bar{z})\mathcal{T}_{\bar{\mathcal{S}}}(\bar{z})$ for tangential cones. As a preparation for the second equality we show that for $\bar{z} \in \partial \bar{\mathcal{S}}$ the matrix

$$\mathrm{D}\Gamma_0(\bar{z}) = A_0 \begin{pmatrix} 1 & 0 & 0 \\ 0 & 1 & 0 \\ \partial_{z_1}\varphi_0 & \partial_{z_2}\varphi_0 & 1 \end{pmatrix} (\bar{z}_1, \bar{z}_2)$$

maps normal vectors on normal vectors in the following way. Recalling the definition of $\nu(z)$ we calculate $\nu^\top(z)\mathrm{D}\Gamma_0(\bar{z}) = \begin{pmatrix} 0 & 0 & -1 \end{pmatrix}\lambda$ for some $\lambda > 0$. Together with the characterization $\mathcal{T}_{\partial\bar{S}}(\bar{z}) = \{v \in \mathbb{R}^3 \ : \ v_3 = 0\}$ and equation (6.1) this proves the second equality (6.2).

We defined the set of constraint forces $\mathcal{N}_{\mathcal{S}}(z) \subset \mathbb{R}^{3^*}$ in Section 2.1.4 via the tangential cone $\mathcal{T}_{\mathcal{S}}(z)$. Using the above formula $\mathcal{T}_{\mathcal{S}}(z) = \mathrm{D}\Gamma_0(\bar{z})\mathcal{T}_{\bar{S}}(\bar{z})$ for the tangential cones and the characterization $\mathcal{T}_{\bar{S}}(\bar{z}) = \{w \in \mathbb{R}^3 \ : \ w_3 \geq 0\}$ we deduce

$$\mathcal{N}_{\mathcal{S}}(z) \stackrel{\text{def}}{=} \left\{v^* \in \mathbb{R}^{3^*} \ : \ v^*v \leq 0 \text{ for all } v \in \mathcal{T}_{\mathcal{S}}(z)\right\}$$
$$= \left\{v^* \in \mathbb{R}^{3^*} \ : \ v^*\mathrm{D}\Gamma_0(\bar{z})w \leq 0 \text{ for all } w \in \mathbb{R}^3, w_3 \geq 0\right\}.$$

Hence for arbitrary $v^* \in \mathcal{N}_{\mathcal{S}}(z)$ we find that $u^* := v^*\mathrm{D}\Gamma_0(\bar{z}) \in \mathbb{R}^{3^*}$ has the form $u^* = \begin{pmatrix} 0 & 0 & -1 \end{pmatrix}\lambda$ for some $\lambda \geq 0$. Again due to the above established relation $\nu^\top(z)\mathrm{D}\Gamma_0(\bar{z}) = \begin{pmatrix} 0 & 0 & -1 \end{pmatrix}\lambda$ we conclude $v^* = \lambda\nu^\top(z)$ for some $\lambda \geq 0$. This proves the third and last equality (6.3).   ∎

6.3 Assumptions, problem and result

In this section we present the assumptions on the data $\mathcal{E}$, $\mathrm{M}^{(j)}$ and especially on the set $\mathcal{S} \subset \mathbb{R}^3$ and its boundary $\partial\mathcal{S}$. The most important assumption will be the new joint convexity Assumptions ((C4) for the general N particle case and the weak Assumption (CO5) for the special one particle case) since these assumptions reveal the influence of the shape of the boundary. Most of the following assumptions are formally equivalent to the assumptions we made in the situation of a flat boundary. To keep this section self-contained we present them here again. We number the assumptions with a the letter 'C' for 'curved'.

The first and new assumption is on the admissible set $\mathcal{S} \subset \mathbb{R}^3$ with $\mathcal{S} \neq \mathbb{R}^3$ and we claim that

$$\mathcal{S} = \text{clos}(\text{int}\mathcal{S}) \quad \text{and} \quad \partial\mathcal{S} \text{ belongs locally to } \mathrm{C}^3. \tag{C0}$$

We denote by $\nu \in \mathrm{C}^2(\partial\mathcal{S}, \mathbb{S}^2)$ the outward normal vector. For the definition of the admissible set $\mathcal{A} \subset \mathbb{R}^{3N}$ for the whole system of N particles we use the same notation as in Chapter 5 and define

$$\mathcal{A} := \left\{z \in \mathbb{R}^{3N} \ : \ z^{(j)} \in \mathcal{S} \quad \text{for } j = 1, \dots, N\right\}.$$

As next data we present the energy functional $\mathcal{E}$ and the typical regularity assumption is

$$\mathcal{E} \in \mathrm{C}^2\big([0, T] \times \mathcal{A}, [0, \infty)\big). \tag{C1}$$

The last data we have to introduce are the matrices of friction and for $j = 1, \ldots, N$ we claim

$$\mathrm{M}^{(j)} \in \mathrm{C}^1\big(\partial \mathcal{S}, \mathbb{R}^{3 \times 3}\big) \quad \text{with} \quad \mathrm{M}^{(j)}(z)\nu(z) = 0 \text{ for all } z \in \partial \mathcal{S}. \tag{C2}$$

After having introduced the data $\mathcal{A}, \mathcal{E}$ and $\mathrm{M}^{(j)}$ the normal force of the j-th particle is as usual given via $\sigma^{(j)}(t, z) := -(\mathrm{D}\mathcal{E})^{(j)}(t, z)\nu(z^{(j)})$ for $t \in [0, T]$ and $z^{(j)} \in \partial \mathcal{S}$ and the dissipation functional $\Psi^{(j)} : [0, T] \times \mathcal{A} \times \mathbb{R}^3 \to [0, \infty)$ for the j-th particle via

$$\Psi^{(j)}(t, z, v) := \begin{cases} \sigma^{(j)}(t, z)_+ \, \|\mathrm{M}^{(j)}(z^{(j)})v\| & \text{for } z^{(j)} \in \partial \mathcal{S}, \\ 0 & \text{else.} \end{cases}$$

The dissipation functional $\Psi : [0, T] \times \mathcal{A} \times \mathbb{R}^{3N} \to [0, \infty)$ for the whole system is then given by

$$\Psi(t, z, v) := \sum_{j=1}^{N} \Psi^{(j)}(t, z, v^{(j)}).$$

The constraint forces are the forces the particles exert on the obstacle. They are the opposite of the forces which the obstacle imposes versus the particle to prevent the particle from penetrating the obstacle. We express them formally using

$$\mathcal{N}_{\mathcal{A}}(z) := \big\{ v^* \in \mathbb{R}^{3N^*} \, : \, v^* v \leq 0 \text{ for all } v \in \mathcal{T}_{\mathcal{A}}(z) \big\}. \tag{6.4}$$

See also Subsection 2.1.4 with the equations (2.6) and (2.7) for the precise definition of $\mathcal{T}_{\mathcal{A}}(z)$. After having clarified the structure of the constraint forces we now present our problem to solve which is formally identical with the previous problems to solve.

Problem 6.4 *For given initial time* $T_0 \in [0, T)$ *and initial state* $Z_0 \in \mathcal{A}$ *find a positive time span* $\Delta \in (0, T{-}T_0]$ *and a solution* $z \in \mathrm{W}^{1,\infty}([T_0, T_0{+}\Delta], \mathcal{A})$ *such that the initial condition* $z(T_0) = Z_0$ *is satisfied and such that for almost all* $t \in [T_0, T_0{+}\Delta]$ *the following differential inclusion holds*

$$0 \in \mathrm{D}\mathcal{E}(t, z(t)) + \partial_v \Psi(t, z(t), \dot{z}(t)) + \mathcal{N}_{\mathcal{A}}(z(t)) \quad \subset \mathbb{R}^{3N^*}. \tag{DI}$$

Obviously we have to assume for the initial state $Z_0 \in \mathcal{A}$ that it satisfies the force balance at time $t = T_0$:

$$0 \in \mathrm{D}\mathcal{E}(T_0, Z_0) + \partial_v \Psi(T_0, Z_0, 0) + \mathcal{N}_{\mathcal{A}}(Z_0) \subset \mathbb{R}^{3N^*}. \tag{C3}$$

If the particle starts on the boundary and hence friction may occur we need a joint convexity assumption for the energy and the dissipation functional. This is the only assumption which is influenced by the curvature or the shape of the boundary. Assume the j-th particle to be in contact with the boundary at time T_0, i.e. $Z_0^{(j)} \in \partial \mathcal{S}$. We denote by $A_j \in \mathbb{R}^{3 \times 3}$ and φ_j the corresponding rotation and shape function which describe the boundary $\partial \mathcal{S}$ in some neighborhood of $Z_0^{(j)}$, see Definition 6.1. Then we describe the effect of the curvature of the boundary on the assumptions via the following matrix

$$B_j := A_j \begin{pmatrix} \partial^2_{z_1^2} \varphi_j(0) & \partial^2_{z_1 z_2} \varphi_j(0) & 0 \\ \partial^2_{z_2 z_1} \varphi_j(0) & \partial^2_{z_2^2} \varphi_j(0) & 0 \\ 0 & 0 & 0 \end{pmatrix} A_j^\top \in \mathbb{R}^{3 \times 3}. \tag{6.5}$$

For the convenience of the reader we recall the definition of the index sets which we introduced in Chapter 5. We have $\mathcal{I} = \{1, \ldots, N\}$, while the contact index set $\mathcal{I}_c$ is given by $\mathcal{I}_c = \{j \in \mathcal{I} : \sigma^{(j)}(T_0, Z_0) > 0\}$ and describes the set of indeces of those particles which are assumed to remain in contact with the obstacle for short times. The switching index set $\mathcal{I}_s = \{j \in \mathcal{I} : \sigma^{(j)}(T_0, Z_0) = 0, Z_0^{(j)} = 0\}$ describes all particles that might switch arbitrary many times within a short time interval between contact with friction and being friction-free. We define for given index set $\mathcal{K} \subset \mathcal{I}$ (the case $\mathcal{K} = \emptyset$ is allowed!) satisfying $\mathcal{I}_c \subset \mathcal{K} \subset \mathcal{I}_c \cup \mathcal{I}_s$ the test set $\mathbb{S}_\mathcal{K}^{3N-1} := \{v \in \mathbb{S}^{3N-1} : v^{(j)} \in \mathcal{T}_{\partial \mathcal{S}}(Z_0^{(j)}), j \in \mathcal{K}\}$ and the constants $\alpha_\mathcal{K} > 0$ and $q_\mathcal{K} > 0$ via

$$\alpha_\mathcal{K} := \min \left\{ v^\top H(T_0, Z_0) v + \sum_{j \in \mathcal{I}_c} \sigma^{(j)}(T_0, Z_0)(v^{(j)})^\top B_j v^{(j)} : v \in \mathbb{S}_\mathcal{K}^{3N-1} \right\}, \tag{6.6}$$

$$q_\mathcal{K} := \max \Big\{ \sum_{j \in \mathcal{K}} |D\sigma^{(j)}(T_0, Z_0)u| \, \|M^{(j)}(Z_0^{(j)})v^{(j)}\| \tag{6.7}$$
$$+ \sigma^{(j)}(T_0, Z_0)_+ \, \|DM^{(j)}(Z_0^{(j)})[v^{(j)}, u^{(j)}]\| : u, v \in \mathbb{S}_\mathcal{K}^{3N-1} \Big\}.$$

Theorem 6.5 (Existence) *Let us assume* (C0)–(C3). *For all index sets* $\mathcal{I}_c \subset \mathcal{K} \subset \mathcal{I}_c \cup \mathcal{I}_s$ *we further assume*

$$\alpha_\mathcal{K} > q_\mathcal{K}. \tag{C4}$$

Then there exists a time span $\Delta > 0$ *and a solution* $z \in W^{1,\infty}([T_0, T_0 + \Delta], \mathcal{A})$ *of Problem 5.1.*

Proof: The proof can be found in Corollary 6.17.   ∎

We now immediately discuss the influence of the curved boundary on the Assumption (C4). The influence of the curved geometry is clear in the elastic term $\alpha_\mathcal{K}$ and is represented by the matrix B_j. But also the frictional terms depend on the same matrix B_j. For this note that $\sigma^{(j)} = (D\mathcal{E})^{(j)} \nu$ and calculating $D\sigma^{(j)}$ leads again to a term including B_j. Since the way B_j influences the question of existence is both times the same we discuss it only for the constant $\alpha_\mathcal{K}$. Of course the shape matters only when the particle is in contact. Note that due to Lemma 6.3 we find

$v^{(j)} \in \mathcal{T}_{\partial S}(Z_0^{(j)})$ to be equivalent to $\bar{v} := A_j^\top v^{(j)} \in \{\bar{v} \in \mathbb{R}^3 : \bar{v}_3 = 0\}$. Hence, the term $(v^{(j)})^\top B_j v^{(j)}$ is equivalent to

$$
\begin{pmatrix} \bar{v}_1 & \bar{v}_2 \end{pmatrix}
\begin{pmatrix} \partial^2_{z_1^2}\varphi_j(0) & \partial^2_{z_1 z_2}\varphi_j(0) \\ \partial^2_{z_2 z_1}\varphi_j(0) & \partial^2_{z_2^2}\varphi_j(0) \end{pmatrix}
\begin{pmatrix} \bar{v}_1 \\ \bar{v}_2 \end{pmatrix}
$$

for some $\bar{v}_1, \bar{v}_2 \in \mathbb{R}$. This shows that a convex shape of the boundary increases $\alpha_\mathcal{K}$ and makes the problem easier to solve while a concave shape of the boundary decreases $\alpha_\mathcal{K}$ and we need a stronger convexity of the energy to compensate the boundary effect. See also Figure 6.3.

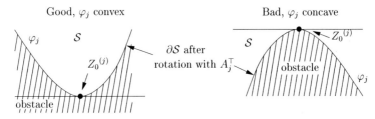

Figure 6.3: Desirable and problematic shape of the boundary

Before we show in the next section how to solve the curved Problem 6.4 we prove a useful component-wise presentation of the set of normal forces.

Lemma 6.6 Let (O0) hold for the set $\mathcal{S} \subset \mathbb{R}^3$ and let the set $\mathcal{A} \subset \mathbb{R}^{3N}$ be defined as above via $\mathcal{A} := \{z \in \mathbb{R}^{3N} : z^{(j)} \in \mathcal{S} \text{ for } j = 1, \ldots, N\}$ then the constraint forces $\mathcal{N}_\mathcal{A}(z) \subset \mathbb{R}^{3N^*}$ can be equivalently defined component-wise. For $j = 1, \ldots, N$ we have

$$(\mathcal{N}_\mathcal{A}(z))^{(j)} = \mathcal{N}_\mathcal{S}(z^{(j)}) \quad \subset \mathbb{R}^{3^*}. \tag{6.8}$$

Proof: We first prove an analog formula for the tangential cone, see (6.9) below. By definition we have $\mathcal{T}_\mathcal{A}(z) = \cap_{\varepsilon>0} \mathcal{T}_\mathcal{A}^\varepsilon(z)$ with

$$\mathcal{T}_\mathcal{A}^\varepsilon(z) = \text{clos}\left\{v \in \mathbb{R}^{3N} : \exists \lambda > 0, \lambda v + z \in \mathcal{A} \cap \mathcal{B}_\varepsilon(z)\right\}.$$

Due to $\mathcal{B}_\varepsilon(z) \subset \{y \in \mathbb{R}^{3N} : y^{(j)} \in \mathcal{B}_\varepsilon(z^{(j)}) \subset \mathbb{R}^3, j = 1, \ldots, N\} \subset \mathcal{B}_{\sqrt{N}\varepsilon}(z)$ we find the relation

$$\mathcal{T}_\mathcal{A}^\varepsilon(z) = \text{clos}\left\{v \in \mathbb{R}^{3N} : \exists \lambda > 0 \text{ with } \lambda v^{(j)} + z^{(j)} \in \mathcal{S} \cap \mathcal{B}_\varepsilon(z^{(j)}) \text{ for } j = 1, \ldots, N\right\}.$$

This proves $(\mathcal{T}_\mathcal{A}^\varepsilon(z))^{(j)} = \mathcal{T}_\mathcal{S}^\varepsilon(z^{(j)})$ from which we deduce

$$(\mathcal{T}_\mathcal{A}(z))^{(j)} = \mathcal{T}_\mathcal{S}(z^{(j)}) \quad \subset \mathbb{R}^3 \tag{6.9}$$

for $j = 1, \ldots, N$. This is the analogon of equation (6.8).

We now start with the proof of (6.8) and first show $(\mathcal{N}_{\mathcal{A}}(z))^{(j)} \subset \mathcal{N}_{\mathcal{S}}(z^{(j)})$. For this we take $v^* \in \mathcal{N}_{\mathcal{A}}(z)$ and prove $(v^*)^{(j)} \in \mathcal{N}_{\mathcal{S}}(z^{(j)})$. For arbitrary $v_j \in \mathcal{T}_{\mathcal{S}}(z^{(j)}) \subset \mathbb{R}^3$ we define the vector $v \in \mathbb{R}^{3N}$ via $v^{(j)} := v_j$ and $v^{(k)} := 0 \in \mathbb{R}^3$ for $k \neq j$. Due to formula (6.9) we have $v \in \mathcal{T}_{\mathcal{A}}(z)$ and we thus find the inequality

$$0 \geq v^* v = (v^*)^{(j)} v^{(j)} = (v^*)^{(j)} v_j.$$

This proves $(v^*)^{(j)} \in \mathcal{N}_{\mathcal{S}}(z)$ since the choice of v_j was arbitrary.

Let us now assume $v_j^* \in \mathcal{N}_{\mathcal{S}}(z^{(j)}) \subset \mathbb{R}^{3^*}$ to hold for all $j = 1, \ldots, N$ and define the vector $v^* := \begin{pmatrix} v_1^* & \ldots & v_N^* \end{pmatrix} \subset \mathbb{R}^{3N^*}$. Our aim is to show $v^* \in \mathcal{N}_{\mathcal{A}}(z)$. For this we assume some $v \in \mathcal{T}_{\mathcal{A}}(z) \subset \mathbb{R}^{3N}$ to be given. Using again formula (6.9) we find $v^{(j)} \in \mathcal{T}_{\mathcal{S}}(z^{(j)}) \subset \mathbb{R}^3$. We thus calculate

$$v^* v = \sum_{j=1}^{N} v_j^* v^{(j)} \leq 0$$

which proves $v^* \in \mathcal{N}_{\mathcal{A}}(z)$. ∎

6.4 Transformation to a problem with flat boundary

The principal idea in the proof of the existence result is the reduction of the curved Problem 6.4 to the flat Problem 5.1. To distinguish the problem formulations we will denote all data and solutions that correspond to the flat situation with a bar '⁻' for flat. For example we define the flat admissible set $\bar{\mathcal{A}}$ via

$$\bar{\mathcal{A}} := \left\{ z \in \mathbb{R}^{3N} \; : \; z_{3j} \geq 0 \quad \text{for all } j = 1, \ldots, N \right\}. \tag{6.10}$$

This flat admissible set now coincides with the admissible set of Chapter 5, see Assumption (N0).

The most important tool will be a flattening transformation $\Phi \in C^2(\bar{\mathcal{A}}, \mathcal{A})$ with which we will put down the curved to the flat problem. The transformation Φ will consist of N single transformations denoted either by Φ_s or $\Phi_j \in C^2(\bar{\mathcal{S}}, \mathcal{S})$ in the following. This single transformations flatten the boundary $\partial \mathcal{S}$ in the neighborhood of the particle $Z_0^{(j)}$. We define next the desired properties for such single transformations.

Definition 6.7 (Single transformation Φ_s) *For given $Z_0 \in \mathcal{S}$ we call a mapping $\Phi_s : \bar{\mathcal{S}} \to \mathcal{S}$ a single transformation if there exists $\bar{Z}_0 \in \bar{\mathcal{S}}$ and a radius $r > 0$ such that the following properties hold:*

1. $\Phi_s \in C^2(\bar{S}, S)$,

2. $\Phi_s : \partial \bar{S} \to \partial S$ and

3. $D\Phi_s(\bar{z}) \begin{pmatrix} 0 & 0 & -1 \end{pmatrix}^\top = \lambda(\bar{z})\nu(\Phi_s(\bar{z}))$ for all $\bar{z} \in \partial \bar{S}$ and $\lambda : \partial S \to (0, 1]$.

We call the three above properties the global properties as they are independent of r. With $\mathcal{U}_r(\bar{Z}_0) := \{\Phi_s(\bar{z}) : \bar{z} \in \mathcal{B}_r(\bar{Z}_0) \cap \bar{S}\}$ the three local properties of Φ_s read

4. $Z_0 = \Phi_s(\bar{Z}_0)$ and Φ_s is a bijection between the sets
   $\mathcal{B}_r(\bar{Z}_0) \cap \partial \bar{S}$ and $\mathcal{U}_r(\bar{Z}_0) \cap \partial S$ and
   $\mathcal{B}_r(\bar{Z}_0) \cap \mathrm{int} \bar{S}$ and $\mathcal{U}_r(\bar{Z}_0) \cap \mathrm{int} S$.

5. The matrix $D\Phi_s(\bar{z}) \in \mathbb{R}^{3\times3}$ is invertible for all $\bar{z} \in \mathcal{B}_r(\bar{Z}_0) \cap \bar{S}$.

6. For all $\bar{z} \in \mathcal{B}_r(\bar{Z}_0) \cap \partial \bar{S}$ we have $D\Phi_s(\bar{z}) \begin{pmatrix} 0 & 0 & -1 \end{pmatrix}^\top = \nu(\Phi_s(\bar{z}))$ or equivalently $\lambda(\bar{z}) = 1$ and we have $D\Phi_s(\bar{z})T_{\partial \bar{S}}(\bar{z}) = T_{\partial S}(\Phi_s(\bar{z}))$.

We prove the existence of single transformations Φ_s in the Propositions 6.12 and 6.13 under the Assumption (C0) of having a C^3-boundary ∂S. But before we want to explain how to use such transformations.

Definition 6.8 (Flattening Transformation Φ) For given $\bar{Z}_0 \in \mathcal{A} \subset \mathbb{R}^{3N}$ and $j = 1, \ldots, N$ let $\Phi_j \in C^2(\bar{S}, S)$ be single transformations that correspond to the single particles $\bar{Z}_0^{(j)} \in S$. Then we call the mapping $\Phi \in C^2(\bar{\mathcal{A}}, \mathcal{A})$ defined via

$$\Phi(\bar{z}) := \begin{pmatrix} \Phi_1(\bar{z}^{(1)}) \\ \cdots \\ \Phi_N(\bar{z}^{(N)}) \end{pmatrix} \in \mathcal{A} \subset \mathbb{R}^{3N}$$

the flattening transformation that corresponds to $\bar{Z}_0 \in \mathcal{A}$.

Transformed "flat" data
In the following we assume a flattening transformation $\Phi \in C^2(\bar{\mathcal{A}}, \mathcal{A})$ to be given. To present all flat data at a glance we repeat the definition of the flat admissible set

$$\bar{\mathcal{A}} = \{\bar{z} \in \mathbb{R}^{3N} : \bar{z}_{3j} \geq 0 \text{ for all } j = 1, \ldots, N\} \tag{6.11}$$
$$= \{\bar{z} \in \mathbb{R}^{3N} : \bar{z}^{(j)} \in \bar{S}, j = 1, \ldots, N\}$$

with the 'flat' set $\bar{S} := \{\bar{z} \in \mathbb{R}^3 : \bar{z}_3 \geq 0\}$. This definition coincides with the Assumption (N0) on the admissible set in Chapter 5. With the help of the flattening transformation Φ we define a new energy functional $\bar{\mathcal{E}}$ on $[0, T] \times \bar{\mathcal{A}}$ by

$$\bar{\mathcal{E}}(t, \bar{z}) := \mathcal{E}(t, \Phi(\bar{z})). \tag{6.12}$$

Due to $\Phi \in C^2$ and together with Assumption (C1), i.e. $\mathcal{E} \in C^2$, we directly find $\bar{\mathcal{E}} \in C^2([0,T] \times \bar{\mathcal{A}}, [0, \infty))$, hence the 'flat' Assumption (N1) holds for the flat data $\bar{\mathcal{E}}$. We also redefine the matrices of friction $M^{(j)}$ with the help of the single transformations $\Phi_j = \Phi^{(j)}$ such that $M^{(j)}$ are defined on the 'flat' boundary $\partial \bar{S} = \mathbb{R}^2 \times 0$. The definition is slightly more complicated and for $\bar{z} \in \partial \bar{S}$ we define

$$\bar{M}^{(j)}(\bar{z}) := M^{(j)}(\Phi_j(\bar{z})) D\Phi_j(\bar{z}) \quad \in \mathbb{R}^{3 \times 3}. \tag{6.13}$$

Here we find the reason why we modeled in the whole thesis friction using matrices $M^{(j)}(\bar{z})$ and not a scalar coefficient $\mu^{(j)}$. We now find that the multiplication with $D\Phi_j(\bar{z}) \in \mathbb{R}^{3 \times 3}$ does not change the structure, we still have matrices. The reason for the multiplication with $D\Phi_j(\bar{z})$ will be the chain-rule formula $\bar{M}^{(j)}(\bar{z}(t))\dot{\bar{z}}(t) = M^{(j)}(z(t))\dot{z}(t)$ for functions $\bar{z}$ and z satisfying the relation $z(t) := \Phi_j(\bar{z}(t))$. Again it is easy to see that (C2) implies the 'flat' Assumption (N2) for the flat friction matrices $\bar{M}^{(j)}$.

Next we introduce the flat dissipation function of the j-th particle $z^{(j)}$ as we did in Chapter 5, too. We define a single dissipation $\bar{\Psi}^{(j)} : [0,T] \times \bar{\mathcal{A}} \times \mathbb{R}^3 \to [0, \infty)$ via

$$\bar{\Psi}^{(j)}(t, \bar{z}, \bar{v}) := \begin{cases} \bar{\sigma}^{(j)}(t, \bar{z})_+ \|\bar{M}^{(j)}(\bar{z}^{(j)})\bar{v}\| & \text{if } \bar{z}^{(j)} \in \partial \bar{S} \\ 0 & \text{else} \end{cases}, \tag{6.14}$$

with the normal force $\bar{\sigma}^{(j)}(t, \bar{z}) := (D\bar{\mathcal{E}})^{(j)}(t, \bar{z}) \begin{pmatrix} 0 & 0 & -1 \end{pmatrix}^{\top}$ and $\|\cdot\|$ being the usual Euclidian norm. The flat dissipation functional $\bar{\Psi} : [0,T] \times \bar{\mathcal{A}} \times \mathbb{R}^{3N} \to [0, \infty)$ for the whole system is now defined as usual via

$$\bar{\Psi}(t, \bar{z}, \bar{v}) := \sum_{j=1}^{N} \bar{\Psi}^{(j)}(t, \bar{z}, \bar{v}^{(j)}).$$

We define the set of constraint forces via

$$\mathcal{N}_{\bar{\mathcal{A}}}(\bar{z}) := \left\{ \bar{v}^* \in \mathbb{R}^{3N*} : \bar{v}^* \bar{v} \leq 0 \text{ for all } \bar{v} \in \mathcal{T}_{\bar{\mathcal{A}}}(\bar{z}) \right\}.$$

or equivalently, see Lemma 6.6, for $j = 1, \ldots, N$ via the formula

$$(\mathcal{N}_{\bar{\mathcal{A}}})^{(j)}(\bar{z}) = \mathcal{N}_{\bar{S}}(\bar{z}^{(j)}) = \begin{cases} \{0\} & \text{if } \bar{z}^{(j)} \in \text{int}\bar{S} \\ \{(0 \quad 0 \quad -\lambda) : \lambda \geq 0\} & \text{if } \bar{z}^{(j)} \in \partial \bar{S}. \end{cases}$$

After having redefined the energy functional $\bar{\mathcal{E}}$ and the dissipational distance $\bar{\Psi}$ we are now able to introduce an auxiliary and flat problem with the new data.

Problem 6.9 (Transformed flat problem) *Let $\bar{\mathcal{A}}, \bar{\mathcal{E}}$ and $\bar{\Psi}$ be defined as above. For given initial time $T_0 \in [0,T)$ find a positive time span $\bar{\Delta} \in (0, T-T_0)$ and a solution $\bar{z} \in W^{1,\infty}([T_0, T_0 + \bar{\Delta}], \bar{\mathcal{A}})$ such that the initial condition $\bar{z}(T_0) = \bar{Z}_0 \in \bar{\mathcal{A}} \subset$*

$\mathbb{R}^{3N}$ *is satisfied and such that for almost all* $t \in [T_0, T_0 + \bar{\Delta}]$ *the following differential inclusion holds*

$$0 \in D\bar{\mathcal{E}}(t, \bar{z}(t)) + \partial_v \bar{\Psi}(t, \bar{z}(t), \dot{\bar{z}}(t)) + \mathcal{N}_{\bar{A}}(\bar{z}(t)) \quad \subset \mathbb{R}^{3N^*}. \tag{$\overline{\text{DI}}$}$$

This flat problem coincides with the N-particles Problem 5.1 of Chapter 5 since the admissible set there coincides with our flat admissible set $\bar{A}$ here. Hence, we directly can solve this flat Problem 6.9 by applying the corresponding existence Theorem 5.2. The following theorem states that this flat problem is equivalent with the desired curved Problem 6.4 which we are interested in.

Theorem 6.10 (Equivalence between flat and curved problems)
Let $\Phi \in C^2(\bar{A}, A)$ *be a flattening transformation then the curved Problem 6.4 and the corresponding flat Problem 6.9 are equivalent. Further the corresponding solutions* $z \in W^{1,\infty}([T_0, T_0+\Delta], A)$ *and* $\bar{z} \in W^{1,\infty}([T_0, T_0+\Delta], \bar{A})$ *can be chosen such that the following formal relations hold:*

$$z = \Phi \circ \bar{z} \quad and \quad \bar{z} = \Phi^{-1} \circ z.$$

Proof: Let $\Phi = \in C^2(\bar{A}, A)$ be a flattening transformation having the usual structure

$$\Phi(z) = \begin{pmatrix} \Phi_1(z^{(1)}) \\ \cdots \\ \Phi_N(z^{(N)}) \end{pmatrix}$$

with single transformations $\Phi_j \in C^2(\bar{S}, S), j = 1, \ldots, N$. We will make strong use of the local Properties (4)–(6) of the single transformations Φ_j, see Definition 6.7 and 6.8. For this we recall that there exists a radius $r > 0$ such that all single transformations Φ_j are invertible on the sets $\mathcal{B}_r(\bar{Z}_0^{(j)}) \cap \bar{S} \subset \mathbb{R}^3$ and the same holds for the matrices $D\Phi_j(\bar{z}) \in \mathbb{R}^{3 \times 3}$ on $\mathcal{B}_r(\bar{Z}_0^{(j)}) \cap \bar{S} \subset \mathbb{R}^3$. Hence, if we restrict the flattening transformation Φ to the set $\mathcal{B}_r(\bar{Z}_0) \cap \bar{A} \subset \mathbb{R}^{3N}$ then we find Φ and $D\Phi \in \mathbb{R}^{3N \times 3N}$ to be invertible. With $\mathcal{U} \subset A$ we denote the image $\mathcal{U} := \{z \in A : z = \Phi(\bar{z}), \bar{z} \in \mathcal{B}_r(\bar{Z}_0) \cap \bar{A}\}$.

After this preliminary observation we assume z to be a given solution of the curved Problem 6.4. We shorten the time span $\Delta > 0$ such that $z \in W^{1,\infty}([T_0, T_0+\Delta], \mathcal{U})$ holds and define the flat solution candidate $\bar{z} \in W^{1,\infty}([T_0, T_0+\Delta], \bar{A} \cap \mathcal{B}_r(\bar{Z}_0))$ via $\bar{z} := \Phi^{-1} \circ z$. We proceed the same way if the flat solution $\bar{z}$ is first given. Hence, as starting point we assume two functions $z \in W^{1,\infty}([T_0, T_0+\Delta], \mathcal{U})$ and $\bar{z} \in W^{1,\infty}([T_0, T_0+\Delta], \bar{A} \cap \mathcal{B}_r(\bar{Z}_0))$ to be given. One of them is a solution and the other a solution candidate of the Problems 6.4 and 6.9. Further they satisfy the relation $z = \Phi \circ \bar{z}$ or $\bar{z} = \Phi^{-1} \circ z$. Hence, we aim to prove both implications at once and do not specify in the following which function is a solution and which function the candidate.

In the following we are interested in the three identities:

$$D\bar{\mathcal{E}}(t, \bar{z}(t)) = D\mathcal{E}(t, z(t))D\Phi(\bar{z}(t)), \tag{6.15}$$

$$\partial_v\bar{\Psi}(t, \bar{z}(t), \dot{\bar{z}}(t)) = \partial_v\Psi(t, z(t), \dot{z}(t))D\Phi(\bar{z}(t)) \quad \text{and} \tag{6.16}$$

$$\mathcal{N}_{\bar{A}}(\bar{z}(t)) = \mathcal{N}_A(z(t))D\Phi(\bar{z}(t)). \tag{6.17}$$

If these equalities hold and $\bar{z}$ is a flat solution then multiplication of the differential inclusion $(\overline{\text{DI}})$ from the right with $(D\Phi)^{-1}(\bar{z}(t))$ shows that z solves (DI) and hence the curved Problem 6.4. If the other way round the curved solution z is first given then multiplication of (DI) from the right with $D\Phi(\Phi^{-1}(z(t)))$ shows that $\bar{z}$ solves $(\overline{\text{DI}})$ and hence the flat Problem 6.9.

Thus it remains us to prove the equalities (6.15)–(6.17). The first equality (6.15) follows directly from the definition of $\bar{\mathcal{E}}$. Due to the structure

$$D\Phi(\bar{z}) = \begin{pmatrix} D\Phi_1(\bar{z}^{(1)}) & 0 & \cdots & 0 \\ 0 & D\Phi_2(\bar{z}^{(2)}) & \cdots & 0 \\ \cdots & \cdots & \cdots & \cdots \\ 0 & 0 & \cdots & D\Phi_N(\bar{z}^{(N)}) \end{pmatrix}$$

the second equality (6.16) is equivalent to have component-wise for all $j = 1, \ldots, N$ the following relation $\partial_v\bar{\Psi}^{(j)}(t, \bar{z}(t), \dot{\bar{z}}^{(j)}(t)) = \partial_v\Psi^{(j)}(t, z(t), \dot{z}^{(j)}(t))D\Phi_j(\bar{z}^{(j)}(t)) \subset \mathbb{R}^{3^*}$. The proof of this formula is established in (6.18) below. First note that since single transformations map normal vectors on normal vectors we have $\bar{\sigma}^{(j)}(t, \bar{z}(t)) = \sigma^{(j)}(t, z(t))$. To keep notation simple we write $\bar{\sigma}^{(j)}$ instead of $\bar{\sigma}^{(j)}(t, \bar{z}(t))$, $\bar{m}(t)$ instead of $\bar{M}^{(j)}(\bar{z}^{(j)})\dot{\bar{z}}^{(j)}(t)$ and $\bar{y}$ instead of $\bar{z}^{(j)}$ in the following formula. We recall the calculus for $\partial_v\bar{\Psi}$, see Theorem 2.3, i.e.

$$\partial_v\bar{\Psi}^{(j)}(t, \bar{y}(t), \dot{\bar{y}}(t)) = \begin{cases} \left\{\bar{\sigma}^{(j)}\left(\frac{\bar{m}(t)}{\|\bar{m}(t)\|}\right)^\top \bar{M}^{(j)}(\bar{y}(t))\right\} & \text{if } \bar{m}(t) \neq 0, \bar{y}(t) \in \partial\bar{\mathcal{S}} \\ \left\{\bar{\sigma}^{(j)}v^*\bar{M}^{(j)}(\bar{y}(t)) : v^* \in \mathcal{B}_1^*(0)\right\} & \text{if } \bar{m}(t) = 0, \bar{y}(t) \in \partial\bar{\mathcal{S}} \\ \{0\} & \text{if } \bar{y}(t) \notin \partial\bar{\mathcal{S}}. \end{cases}$$

The following step now elucidates why we modeled the dissipation relatively general by introducing matrices of friction $M^{(j)}(z) \in \mathbb{R}^{3\times3}$. This allowed us to define the friction in the flat case by multiplication with the matrix $D\Phi_j(\bar{z}) \in \mathbb{R}^{3\times3}$ from the right, i.e. $\bar{M}^{(j)}(\bar{y}) = M^{(j)}(\Phi_j(\bar{y}))F_j$ with $F_j = D\Phi_j(\bar{y})$. Due to $\dot{z}^{(j)}(t) = F_j\dot{\bar{z}}^{(j)}(t)$ this provides us with the desired formula

$$\bar{M}^{(j)}(\bar{y}(t))\dot{\bar{y}}(t) = \bar{M}^{(j)}(\bar{z}^{(j)}(t))\dot{\bar{z}}^{(j)}(t) = M^{(j)}(z^{(j)}(t))\dot{z}^{(j)}(t) = M^{(j)}(y(t))\dot{y}(t)$$

Hence we rewrite the above formula for the subdifferential $\partial_v\bar{\Psi}^{(j)}$ as

$$\partial_v \bar{\Psi}^{(j)}(t, \bar{y}(t), \dot{\bar{y}}(t)) = \begin{cases} \left\{ \sigma^{(j)} \left(\frac{m(t)}{\|m(t)\|} \right)^\top \mathrm{M}^{(j)}(y(t)) F_j \right\} & \text{if} \ \ m(t) \neq 0, y(t) \in \partial \mathcal{S} \\ \left\{ \sigma^{(j)} v^* \mathrm{M}^{(j)}(y(t)) F_j : v^* \in \mathcal{B}_1^*(0) \right\} & \text{if} \ \ m(t) = 0, y(t) \in \partial \mathcal{S} \\ \{0\} & \text{if} \ \ y(t) \notin \partial \mathcal{S}. \end{cases}$$

Here we used that Φ_j is locally a bijection between the boundaries $\bar{\mathcal{S}}$ and $\partial \mathcal{S}$. This proves

$$\partial_v \bar{\Psi}^{(j)}(t, \bar{z}(t), \dot{\bar{z}}^{(j)}(t)) = \partial_v \Psi^{(j)}(t, z(t), \dot{z}^{(j)}(t)) \mathrm{D}\Phi_j(\bar{z}^{(j)}(t)) \subset \mathbb{R}^{3^*}. \tag{6.18}$$

Hence, (6.16) is established component-wise.

In the same spirit as for the second identity the third identity (6.17) is equivalent to

$$\mathcal{N}_{\bar{\mathcal{S}}}(\bar{z}^{(j)}) = \mathcal{N}_{\mathcal{S}}(z^{(j)}(t)) \mathrm{D}\Phi_j(\bar{z}^{(j)}(t)) \quad \text{for all } j = 1, \ldots, N.$$

The equivalence is a consequence of Lemma 6.6 where we have shown $(\mathcal{N}_{\bar{A}})^{(j)}(\bar{z}) = \mathcal{N}_{\bar{\mathcal{S}}}(\bar{z}^{(j)})$. For $\bar{z}^{(j)}(t) \in \mathrm{int}\bar{\mathcal{S}}$ or equivalently $z^{(j)}(t) \in \mathrm{int}\mathcal{S}$ we have $\mathcal{N}_{\bar{\mathcal{S}}}(\bar{z}^{(j)}(t)) = \mathcal{N}_{\mathcal{S}}(z^{(j)}(t)) = 0 \in \mathbb{R}^{3^*}$. Hence we consider only the situation of $\bar{z}^{(j)}(t) \in \partial \bar{\mathcal{S}}$ or equivalently $z^{(j)}(t) \in \partial \mathcal{S}$. For $\mathcal{N}_{\bar{\mathcal{S}}}(\bar{z}^{(j)}(t))$ we directly find

$$\mathcal{N}_{\bar{\mathcal{S}}}(\bar{z}^{(j)}(t)) = \left\{ v^* \in \mathbb{R}^{3^*} : v^* = \lambda \begin{pmatrix} 0 & 0 & -1 \end{pmatrix}, \lambda \geq 0 \right\}$$

while we have shown for $z^{(j)}(t)) \in \partial \mathcal{S}$ in Lemma 6.3 the relation

$$\mathcal{N}_{\mathcal{S}}(z^{(j)}(t)) = \left\{ v^* \in \mathbb{R}^{3^*} : v^* = \lambda \nu^\top(z^{(j)}(t)), \lambda \geq 0 \right\}.$$

We thus establish the equality $\mathcal{N}_{\bar{\mathcal{S}}}(\bar{z}^{(j)}(t)) = \mathcal{N}_{\mathcal{S}}(z(t)) \mathrm{D}\Phi_j(\bar{z}^{(j)}(t))$ if we prove

$$\begin{pmatrix} 0 & 0 & -1 \end{pmatrix} = \nu^\top(z^{(j)}(t)) \mathrm{D}\Phi_j(\bar{z}^{(j)}(t)). \tag{6.19}$$

We check the third component of the right expression by multiplying with the vector $\begin{pmatrix} 0 & 0 & 1 \end{pmatrix}^\top$ from the right. This gives $-\nu^\top(z^{(j)}(t))\nu(z^{(j)}(t)) = -1$ since $\mathrm{D}\Phi_j(\bar{z}^{(j)})$ maps normal vectors on normal vectors, see the sixth property of single transformations in Definition 6.7. Also due to its definition the single transformation satisfies $\mathrm{D}\Phi_j(\bar{z}^{(j)})\mathcal{T}_{\partial \bar{\mathcal{S}}}(\bar{z}^{(j)}) = \mathcal{T}_{\partial \mathcal{S}}(\Phi_j(\bar{z}^{(j)}))$. Thus, multiplication of $\nu^\top(z^{(j)}(t))\mathrm{D}\Phi_j(\bar{z}^{(j)}(t))$ with either $\begin{pmatrix} 1 & 0 & 0 \end{pmatrix}^\top$ or $\begin{pmatrix} 0 & 1 & 0 \end{pmatrix}^\top \in \mathcal{T}_{\partial \bar{\mathcal{S}}}(\bar{z}^{(j)}(t))$ from the right gives 0 for the first and second component, since the normal vector ν and the tangential plane $\mathcal{T}_{\partial \mathcal{S}}$ are perpendicular to each other, see Lemma 6.3. This proves (6.19) and hence the third and last equality (6.17). We thus established the last of the three equalities (6.15)–(6.17). This completes our proof. ∎

Corollary 6.11 (The flat Assumptions (N0)–(N3)) *If $\Phi \in C^2(\bar{\mathcal{A}}, \mathcal{A})$ is a flattening transformation then the curved Assumptions (C0)–(C3) on the curved data $\mathcal{S}, \mathcal{E}, M^{(j)}$ and Z_0 imply the flat Assumptions (N0)–(N3) on the flat data $\bar{\mathcal{S}}, \bar{\mathcal{E}}, \bar{M}^{(j)}$ and $\bar{Z}_0$.*

Proof: During the introduction of the flat data we already have shown that (C0)–(C2) imply (N0)–(N2). Using the above Theorem 6.10 we now establish the equivalence of the initial Assumptions (C3) and (N3). For this we define an auxiliary energy functional E which is constant in time via $E(t, z) := \mathcal{E}(T_0, z)$ for all $t \in [T_0, T]$ and $z \in \mathcal{A}$. The corresponding curved problem has the solution $z \equiv Z_0$ which satisfies (C3). The same holds for the corresponding flat problem which has the solution $\bar{z} \equiv \bar{Z}_0$ satisfying (O3). The above theorem then proves the equivalence of (C3) and (N3). ∎

For the equivalence of the convexity Assumptions (C4) and (N4) the abstract Definition 6.8 of a flattening transformation is not sufficient. We now present the concrete constructions of single transformations. The following proposition states the existence of a single transformation Φ_s if a single particle $z_0 \in \mathcal{S}$ has no contact with the boundary. The reader might wonder why we construct a complicated transformation if the boundary does not matter and hence we are directly in the situation of the previous section. We decided to pay this prize since we preferred to have a uniform characterization of all single particles of the system independent of their positions via single transformations. Hence, also the following transformation has to map boundary on boundary, etc..

Proposition 6.12 (Existence of a single transformation for $z_0 \in \mathrm{int}\mathcal{S}$)
We assume $z_0 \in \mathrm{int}\mathcal{S}$ and (C0). Then there exists a mapping $\Phi_s \in C^2(\bar{\mathcal{S}}, \mathcal{S})$ together with a radius $r > 0$ and an initial value $\bar{z}_0 := \begin{pmatrix} 0 & 0 & 2r \end{pmatrix}^\top \in \bar{\mathcal{S}}$ all together satisfying the Definition 6.7 of a single transformation. Further the single transformation Φ_s satisfies

$$D\Phi_s(\bar{z}_0) \in \mathbb{R}^{3\times 3} \quad \text{is orthonormal and}$$
$$D^2\Phi_s(\bar{z}_0) = 0 \in \mathbb{R}^{3\times 3\times 3}.$$

Proof: We define the radius $r := \frac{1}{2}\sup\{\rho \geq 0 \;:\; \mathcal{B}_\rho(z_0) \subset \mathcal{S}\}$. The factor $1/2$ will get clear later on. Due to $z_0 \in \mathrm{int}\mathcal{S}$ and $\mathcal{S} \neq \mathbb{R}^3$ we have $r \in (0, \infty)$ and there exists $z_\mathrm{b} \in \partial\mathcal{S}$, 'b' stands for boundary such that $\|z_0 - z_\mathrm{b}\| = 2r$ and $\frac{1}{2r}(z_0 - z_\mathrm{b}) = -\nu(z_\mathrm{b})$. As stated in the proposition we define $\bar{z}_0 := \begin{pmatrix} 0 & 0 & 2r \end{pmatrix}^\top \in \bar{\mathcal{A}}$ and choose an orthonormal matrix $A \in \mathbb{R}^{3\times 3}$ which rotates normal vector onto normal vector, i.e. $A\bar{z}_0 = z_0 - z_\mathrm{b}$. We now define for $\bar{z} \in \mathbb{R}^3$ the local transformation

$$\Phi_\mathrm{loc}(\bar{z}) := A(\bar{z} - \bar{z}_0) + z_0. \tag{6.20}$$

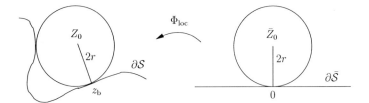

Figure 6.4: The local transformation Φ_{loc}

See also Figure 6.4. We call the transformation local since it satisfies $\Phi_{\text{loc}}(\bar{z}) \in \mathcal{S}$ for all $\bar{z} \in \mathcal{B}_{2r}(\bar{z}_0)$ and $\Phi_{\text{loc}}(\bar{z}_0) = z_0$. Further Φ_{loc} maps the boundary element $0 \in \partial \bar{\mathcal{S}}$ onto the boundary element $z_{\text{b}} \in \partial \mathcal{S}$ and $\mathrm{D}\Phi_{\text{loc}}(0) \begin{pmatrix} 0 & 0 & -1 \end{pmatrix}^{\top} = \nu(z_b)$ holds.
After having introduced a local transformation we introduce introduce a retraction mapping $\Phi_{\text{re}} \in \mathrm{C}^2(\bar{\mathcal{S}}, \mathcal{B}_{2r}(\bar{z}_0))$ and construct Φ_s via $\Phi_s := \Phi_{\text{loc}} \circ \Phi_{\text{re}}$. For this we choose two cutoff functions $f, \eta \in \mathrm{C}^\infty(\mathbb{R}, [0,1])$ with

$$f(x) = 0 \text{ for } x \leq \frac{1}{4}r \text{ and } f(x) = 1 \text{ for } x \geq \frac{1}{2}r \text{ and}$$

$$\eta(x) = 1 \text{ for } x \leq r \text{ and } \eta(x) = 0 \text{ for } x \geq \frac{3}{2}r.$$

The retraction mapping $\Phi_{\text{re}} \in \mathrm{C}^\infty(\mathbb{R}^3, \mathbb{R}^3)$ is then defined for $\bar{z} \in \mathbb{R}^3$ via

$$\Phi_{\text{re}}(\bar{z}) := \begin{pmatrix} \bar{z}_1 f(\bar{z}_3) \eta(\|\bar{z} - \bar{z}_0\|) \\ \bar{z}_2 f(\bar{z}_3) \eta(\|\bar{z} - \bar{z}_0\|) \\ \bar{z}_3 \big(1 - f(\bar{z}_3) + f(\bar{z}_3) \eta(\|\bar{z} - \bar{z}_0\|)\big) \end{pmatrix}.$$

In Figure 6.5 we present the different domains and the resulting definition of the retraction mapping Φ_{re}. The different domains are separated by dotted lines while the set $\mathcal{B}_{2r}(\bar{z}_0)$ is limited by a full line. As a first result note that Φ_{re} retracts the flat admissible set $\bar{\mathcal{S}}$ onto $\mathcal{B}_{2r}(\bar{z}_0)$. Together with the composition $\Phi_s = \Phi_{\text{loc}} \circ \Phi_{\text{re}}$ this proves $\Phi_s \in \mathrm{C}^2(\bar{\mathcal{S}}, \mathcal{S})$. This is the first property of a single transformation. The crucial domains are (I) and (V). The definition of Φ_{re} on the first domain $\{\bar{z} \in \mathbb{R}^3 : \bar{z}_3 \leq \frac{1}{4}r\}$ shows for all $\bar{z} \in \partial \bar{\mathcal{S}}$

$$\Phi_{\text{re}}(\bar{z}) = 0 \text{ and } \mathrm{D}\Phi_{\text{re}}(\bar{z}) \begin{pmatrix} 0 \\ 0 \\ -1 \end{pmatrix} = \begin{pmatrix} 0 \\ 0 \\ -1 \end{pmatrix}.$$

This proves the second and third property of a single transformation, i.e. $\Phi_s(\bar{z}) \in \partial \mathcal{S}$ and $\mathrm{D}\Phi_s(\bar{z}) \begin{pmatrix} 0 & 0 & -1 \end{pmatrix}^{\top} = \nu(\Phi_s(\bar{z}))$ for all $\bar{z} \in \partial \bar{\mathcal{S}}$. The definition of Φ_{re} as the

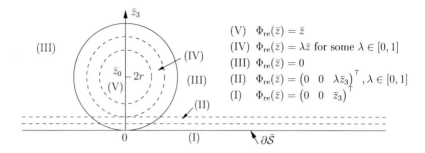

$$(V) \quad \Phi_{re}(\bar{z}) = \bar{z}$$
$$(IV) \quad \Phi_{re}(\bar{z}) = \lambda\bar{z} \text{ for some } \lambda \in [0,1]$$
$$(III) \quad \Phi_{re}(\bar{z}) = 0$$
$$(II) \quad \Phi_{re}(\bar{z}) = \begin{pmatrix} 0 & 0 & \lambda\bar{z}_3 \end{pmatrix}^\top, \lambda \in [0,1]$$
$$(I) \quad \Phi_{re}(\bar{z}) = \begin{pmatrix} 0 & 0 & \bar{z}_3 \end{pmatrix}^\top$$

Figure 6.5: Domains and definitions of Φ_{re}

identity on $\mathcal{B}_r(\bar{z}_0)$ together with the affine structure of Φ_{loc} prove on the one hand the local properties (4)–(6) of Definition 6.7, note that $\partial\bar{\mathcal{S}} \cap \mathcal{B}_r(\bar{z}_0) = \emptyset$. On the other hand this proves the additional properties $D\Phi_s(\bar{z}_0) \in \mathbb{R}^{3\times3}$ being orthonormal and $D^2\Phi_s(\bar{z}_0) = 0$. ∎

For the following proposition we make the calculus convention $(D^2\Phi(z)[u,v])_i = \sum_{j,k=1}^3 \partial^2_{z_j,z_k}\Phi_i(z)u_jv_k$ for $i = 1,\ldots,3$ and $u, v \in \mathbb{R}^3$.

Proposition 6.13 (Existence of a single transformation for $z_0 \in \partial\mathcal{S}$)
We assume $z_0 \in \partial\mathcal{S}$ and (C0). Then there exists a mapping $\Phi_s \in C^2(\bar{\mathcal{S}}, \mathcal{S})$ together with a radius $r > 0$ with $\bar{z}_0 := 0 \in \partial\bar{\mathcal{S}}$ satisfying the Definition 6.7 of a single transformation. Further the single transformation Φ_s satisfies

$$D\Phi_s(0) = A_0 \in \mathbb{R}^{3\times3} \quad and \tag{6.21}$$

$$D^2\Phi_s(0)[\bar{v},\bar{u}] = -(\bar{v}^\top \begin{pmatrix} \partial^2_{z_1^2}\varphi_0(0) & \partial^2_{z_1z_2}\varphi_0(0) & 0 \\ \partial^2_{z_2z_1}\varphi_0(0) & \partial^2_{z_2^2}\varphi_0(0) & 0 \\ 0 & 0 & 0 \end{pmatrix} \bar{u})\nu(Z_0) \quad \in \mathbb{R}^3 \tag{6.22}$$

for all $\bar{u}, \bar{v} \in \mathbb{R}^3$ with $\bar{u}_3 = \bar{v}_3 = 0$ and the rotation A_0 and shape function φ_0 as in Definition 6.1.

Proof: The construction of a single transformation Φ_s, see Definition 6.7, is again done by a concatenation $\Phi_s := \Phi_{loc} \circ \Phi_{re}$ of a local transformation Φ_{loc} and a retraction mapping Φ_{re}. We first present desired properties for Φ_{loc} and Φ_{re} and show that these properties are sufficient to assure $\Phi_s = \Phi_{loc} \circ \Phi_{re}$ to be a single transformation. Only afterwards, at the end of the proof, we give the precise construction of Φ_{loc} and Φ_{re}.

The idea is to choose a local transformation Φ_{loc} which satisfies for some given $r > 0$ the global properties (1)–(3) only locally on a Ball $\mathcal{B}_{2r}(0)$ while the local properties

(4)-(6) are left unchanged. For this see Definition 6.7 of a single transformation Φ_s. Hence, we want Φ_{loc} to satisfy the following properties.

1. $\Phi_{\text{loc}} \in C^2(\bar{\mathcal{S}} \cap \mathcal{B}_{2r}(0), \mathcal{S})$,

2. $\Phi_{\text{loc}} : \partial\bar{\mathcal{S}} \cap \mathcal{B}_{2r}(0) \to \partial\mathcal{S}$ and

3. $\text{D}\Phi_{\text{loc}}(\bar{z}) \begin{pmatrix} 0 & 0 & -1 \end{pmatrix}^\top = \nu(\Phi_{\text{loc}}(\bar{z}))$ for all $\bar{z} \in \partial\bar{\mathcal{S}} \cap \mathcal{B}_{2r}(0)$.

We define the set $\mathcal{U}_r(Z_0) := \left\{ z = \Phi_{\text{loc}}(\bar{z}) \in \mathcal{S} : \bar{z} \in \bar{\mathcal{S}} \cap \mathcal{B}_r(0) \right\}$ for the following properties. They exactly coincide with the three local properties of a single transformation.

4. $Z_0 = \Phi_{\text{loc}}(0)$ and Φ_{loc} is a bijection between the sets
   $\mathcal{B}_r(0) \cap \partial\bar{\mathcal{S}}$ and $\mathcal{U}_r(Z_0) \cap \partial\mathcal{S}$ and
   $\mathcal{B}_r(0) \cap \text{int}\bar{\mathcal{S}}$ and $\mathcal{U}_r(Z_0) \cap \text{int}\mathcal{S}$.

5. The matrix $\text{D}\Phi_{\text{loc}}(\bar{z}) \in \mathbb{R}^{3\times3}$ is invertible for all $\bar{z} \in \bar{\mathcal{S}} \cap \mathcal{B}_r(0)$.

6. We have $\text{D}\Phi_{\text{loc}}(\bar{z})\mathcal{T}_{\bar{\mathcal{S}}}(\bar{z}) = \mathcal{T}_{\mathcal{S}}(\Phi_{\text{loc}}(\bar{z}))$ and $\text{D}\Phi_{\text{loc}}(\bar{z}) \begin{pmatrix} 0 & 0 & -1 \end{pmatrix}^\top = \nu(\Phi_{\text{loc}}(\bar{z}))$ for all $\bar{z} \in \partial\bar{\mathcal{S}} \cap \mathcal{B}_r(0)$.

We next present the required properties for the retraction mapping Φ_{re} whose task it is to expand the definition of Φ_s to the whole set $\bar{\mathcal{S}}$ and to leave at the same time Φ_{loc} unchanged on the inner ball $\mathcal{B}_r(0)$. Hence, we want for Φ_{re} the properties:

1. $\Phi_{\text{re}} \in C^2(\bar{\mathcal{S}}, \bar{\mathcal{S}} \cap \mathcal{B}_{2r}(0))$,

2. $\Phi_{\text{re}} : \partial\bar{\mathcal{S}} \to \partial\bar{\mathcal{S}} \cap \mathcal{B}_{2r}(0)$,

3. $\text{D}\Phi_{\text{re}}(\bar{z}) \begin{pmatrix} 0 & 0 & -1 \end{pmatrix}^\top = \lambda(\bar{z}) \begin{pmatrix} 0 & 0 & -1 \end{pmatrix}^\top$ for all $\bar{z} \in \partial\bar{\mathcal{S}}$ and with $\lambda(\bar{z}) \in (0, 1]$ and

4. $\Phi_{\text{re}}(\bar{z}) = \bar{z}$ for all $\bar{z} \in \bar{\mathcal{S}} \cap \mathcal{B}_r(0)$.

We now directly see that $\Phi_s = \Phi_{\text{loc}} \circ \Phi_{\text{re}}$ satisfies the three global properties $\Phi_s \in C^2(\bar{\mathcal{S}}, \mathcal{S})$, $\Phi_s : \partial\bar{\mathcal{S}} \to \partial\mathcal{S}$ and $\text{D}\Phi_s(\bar{z}) \begin{pmatrix} 0 & 0 & -1 \end{pmatrix}^\top = \lambda(\bar{z})\nu(\Phi_s(\bar{z}))$ for some $\lambda(\bar{z}) \in (0, 1]$ and all $\bar{z} \in \partial\bar{\mathcal{S}}$. The identity of Φ_{re} on $\bar{\mathcal{S}} \cap \mathcal{B}_r(0)$ shows directly that $\Phi_s \equiv \Phi_{\text{loc}}$ satisfies on $\mathcal{B}_r(0)$ the three local properties (4)-(6) of Definition 6.7. It remains to construct Φ_{loc} and Φ_{re}.

We now construct the retraction mapping Φ_{re} for given radius $r > 0$. We make a radial symmetric ansatz $\Phi_{\text{re}}(\bar{z}) := \bar{z}h(\|\bar{z}\|)$ with some strictly positive scalar function $h \in C^2([0, \infty), (0, 1])$. Hence Φ_{re} maps spheres with radius $\|\bar{z}\|$ on spheres with radius $\|\bar{z}\|h(\|\bar{z}\|) = \|\Phi_{\text{re}}(\bar{z})\|$. The idea is to choose this mapping as in Figure 6.6, thus leaving spheres with radius less than r unchanged and retracting all other

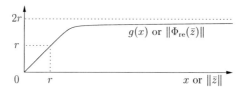

Figure 6.6: Graph of the function g

spheres into the ball $\mathcal{B}_{2r}(0)$. For this we define the function $f \in C^2([0,\infty),[0,4])$ via

$$f(x) := \begin{cases} x & x \le 2, \\ 2 + \frac{x-2}{1+(x-2)} + \frac{(x-2)^2}{1+(x-2)^2} & 2 < x \end{cases}$$

and rescale it via $g(x) := \frac{r}{2}f(2x/r)$. The graph of g is presented in Figure 6.6 and we see that we would like $\|\Phi_{\mathrm{re}}(\bar{z})\| = g(\|z\|)$ to hold. This motivates the choice $h(0) := 1$ and $h(x) := g(x)/x$ for $x > 0$. It is now easy to check that $\Phi_{\mathrm{re}}(\bar{z}) = \bar{z}h(\|\bar{z}\|)$ satisfies the four above required properties for a retraction mapping with the choice $\lambda(\bar{z}) = h(\|\bar{z}\|) \in (0,1]$ in the third property.

We next construct the local transformation Φ_{loc} and fix $r > 0$. The definition of Φ_{loc} is based on the local representation $\Gamma_0 \in C^3(\mathcal{B}_{r_0}(0),\mathbb{R}^3)$ of the boundary $\partial\mathcal{S}$ in a neighborhood of $Z_0 \in \partial\mathcal{S}$, see Definition 6.1. For $\bar{z} \in \mathcal{B}_{r_0}(0) \subset \mathbb{R}^3$ we define $\Phi_{\mathrm{loc}} \in C^2(\mathcal{B}_{r_0}(\bar{z}_0),\mathbb{R}^3)$ via

$$\Phi_{\mathrm{loc}}(\bar{z}) := \Gamma_0(\bar{z}_1, \bar{z}_2, 0) - \bar{z}_3\nu\big(\Gamma_0(\bar{z}_1, \bar{z}_2, 0)\big).$$

See also Figure 6.7. Recalling the precise definition of the local representation Γ_0

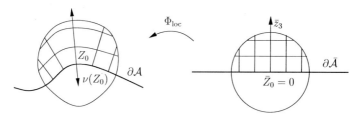

Figure 6.7: Local transformation Φ_{loc}

and the outward normal vector ν in Definition 6.1 we write in full detail

$$\Phi_{\mathrm{loc}}(\bar{z}) := Z_0 + A_0 \begin{pmatrix} \bar{z}_1 \\ \bar{z}_2 \\ \varphi_0(\bar{z}_1, \bar{z}_2) \end{pmatrix} + \frac{\bar{z}_3}{\|(\partial_{z_1}\varphi_0, \partial_{z_2}\varphi_0, -1)(\bar{z}_1, \bar{z}_2)\|}A_0 \begin{pmatrix} -\partial_{z_1}\varphi_0(\bar{z}_1, \bar{z}_2) \\ -\partial_{z_2}\varphi_0(\bar{z}_1, \bar{z}_2) \\ 1 \end{pmatrix}.$$

Note that the rotation $A_0 \in \mathbb{R}^{3\times 3}$ is orthonormal and for the shape function φ_0 we need $\varphi_0 \in C^3(\mathcal{B}_{r_0}(0), \mathbb{R})$ to assure Φ_{loc} to be twice continuously differentiable. This is why we assume a C^3-boundary in the Assumption (C0). To fix the radius $r > 0$ we calculate, for $\bar{z}$ with $\bar{z}_3 = 0$ or equivalently for $\bar{z} \in \partial\bar{\mathcal{S}}$, the derivative

$$\mathrm{D}\Phi_{\text{loc}}(\bar{z}) = A_0 \begin{pmatrix} 1 & 0 & -\psi\partial_{z_1}\varphi_0 \\ 0 & 1 & -\psi\partial_{z_2}\varphi_0 \\ \partial_{z_1}\varphi_0 & \partial_{z_2}\varphi_0 & \psi \end{pmatrix} (\bar{z}_1, \bar{z}_2) \qquad (6.23)$$

with the function $\psi(\bar{z}_1, \bar{z}_2) := 1/\|(\partial_{z_1}\varphi_0, \partial_{z_2}\varphi_0, -1)(\bar{z}_1, \bar{z}_2)\|$. Due to $\partial_{z_1}\varphi(0) = \partial_{z_2}\varphi(0) = 0$, see Definition 6.1, we have $\mathrm{D}\Phi_{\text{loc}}(0) = A_0$. This is (6.21) in our proposition to prove. By a classical result of analysis and since $\mathrm{D}\Phi_{\text{loc}}$ is continuous there exists a radius $0 < r \leq r_0/2$ such that for all $\bar{z} \in \mathcal{B}_{2r}(0) \subset \mathcal{B}_{r_0}(0)$ the matrix $\mathrm{D}\Phi_{\text{loc}}(\bar{z})$ is invertible and Φ_{loc} is a bijection between $\mathcal{B}_{2r}(0)$ and its image $\mathcal{U}_{2r} := \{z \in \mathbb{R}^3 : z = \Phi_{\text{loc}}(\bar{z}), \bar{z} \in \mathcal{B}_{2r}(0)\}$. After having defined Φ_{loc} and $r > 0$ we next verify the six properties which we required above. They directly follow from formula (6.23) and the following argument. We find that Φ_{loc} is also a bijection on the three following disjoint subsets: the boundaries $\partial\bar{\mathcal{S}} \cap \mathcal{B}_{2r}(0)$ and $\partial\mathcal{S} \cap \mathcal{U}_{2r}$, the interior $\text{int}\bar{\mathcal{S}} \cap \mathcal{B}_{2r}(0)$ and $\text{int}\mathcal{S} \cap \mathcal{U}_{2r}$ and the exterior $\bar{\mathcal{S}}^c \cap \mathcal{B}_{2r}(0)$ and $\mathcal{S}^c \cap \mathcal{U}_{2r}$. This holds since on the one hand Φ_{loc} coincides on the boundary $\partial\bar{\mathcal{S}}$ with Γ_0 which is by definition a bijection between the boundaries. On the other hand Φ_{loc} is continuous and hence maps connected sets on connected sets. Since it coincides on the normal axis with Γ_0, i.e. $\Phi_{\text{loc}}(0, 0, \bar{z}_3) = \Gamma_0(0, 0, \bar{z}_3)$ we find the bijection on the interior and exterior sets. For the sixth required property, i.e. $\mathrm{D}\Phi_{\text{loc}}(\bar{z})\mathcal{T}_{\bar{\mathcal{S}}}(\bar{z}) = \mathcal{T}_{\mathcal{S}}(\Phi_{\text{loc}}(\bar{z}))$ and $\mathrm{D}\Phi_{\text{loc}}(\bar{z})\begin{pmatrix} 0 & 0 & -1 \end{pmatrix}^\top = \nu(\Phi_{\text{loc}}(\bar{z}))$, note that in (6.23) the third column of $\mathrm{D}\Phi_{\text{loc}}$ is perpendicular to the first and second.

In remains us to prove the formula (6.22) for the second derivative $\mathrm{D}^2\Phi_s(0)$. We start by splitting for $\bar{z} \in \mathcal{B}_r(0)$ the function $\Phi_s \equiv \Phi_{\text{loc}}$ into four terms

$$\Phi_s(\bar{z}) = \Phi_{\text{loc}}(\bar{z}) = Z_0 + A_0 \begin{pmatrix} \bar{z}_1 \\ \bar{z}_2 \\ 0 \end{pmatrix} + A_0 \begin{pmatrix} 0 \\ 0 \\ \varphi_0(\bar{z}_1, \bar{z}_2) \end{pmatrix} - \bar{z}_3\nu(\Gamma_0(\bar{z}_1, \bar{z}_2, 0)). \quad (6.24)$$

Due to $\bar{v}_3 = \bar{u}_3 = 0$ and by linearity in $\bar{z}_3$ the following formula contains no ∂_{z_3}:

$$\mathrm{D}^2\Phi_s(0)[\bar{v}, \bar{u}] = \mathrm{D}^2\Phi_{\text{loc}}(0)[\bar{v}, \bar{u}] = \sum_{j,k=1}^{2} \partial_{z_k}\partial_{z_j}\Phi_{\text{loc}}(0)\bar{v}_j\bar{u}_k.$$

Obviously the second derivatives of the first, second and forth term in equation (6.24) are zero. For the fourth term we used $\bar{z}_3 = 0$. The remaining third term will supply us with the desired formula (6.22) and we recall $A_0 = \mathrm{D}\Phi_s(0)$ and

$\mathrm{D}\Phi_s(0)(0,0,-1)^\top = \nu(Z_0)$ with ν being the outward normal vector. Hence the third term satisfies

$$A_0 \begin{pmatrix} 0 \\ 0 \\ \varphi_0(\bar{z}_1, \bar{z}_2) \end{pmatrix} = -\varphi_0(\bar{z}_1, \bar{z}_2)\nu(Z_0) \in \mathbb{R}^3.$$

we now have

$$\mathrm{D}^2\Phi_s(0)[\bar{v}, \bar{u}] = -\left(\sum_{j,k=1}^2 \partial_{z_k} \partial_{z_j} \varphi_0(0) \bar{v}_j \bar{u}_k \right) \nu(Z_0).$$

This proves our result desired result (6.22).   ∎

6.5 Retranslating the convexity assumption

In this section we prove the equivalence of the Assumption (N4) and (C4). In Corollary 6.11 we showed that the curved Assumptions (C0)–(C3) imply the flat Assumptions (N0)–(N3). In the following we will show the equivalence between the convexity Assumptions (C4) and (N4). This is the missing piece in the puzzle since having all Assumptions (N0)–(N4) at hand we are able to apply the Existence Theorem 5.2 and thus to solve the flat or equivalently curved Problem 6.4. For the convenience of the reader we repeat here the Assumption (N4) for our flat data. Depending on the initial values T_0 and $\bar{Z}_0 \in \bar{\mathcal{A}}$ we define the index sets:

$$\mathcal{I} := \{1, \ldots, N\},$$
$$\mathcal{I}_c := \left\{ j \in \mathcal{I} \ : \ \bar{\sigma}^{(j)}(T_0, \bar{Z}_0) > 0 \right\},$$
$$\mathcal{I}_f := \left\{ j \in \mathcal{I} \ : \ \left(\bar{Z}_0\right)_{3j} > 0 \right\} \quad \text{and}$$
$$\mathcal{I}_s := \mathcal{I} \backslash \left(\mathcal{I}_c \cup \mathcal{I}_f\right).$$

For given index set $\mathcal{K} \subset \mathcal{I}$ (the case $\mathcal{K} = \emptyset$ is allowed!) we define the set $\bar{\mathbb{S}}_\mathcal{K}^{3N-1} := \left\{ v \in \mathbb{S}^{3N-1} \ : \ v^j \in T_{\partial \bar{S}}(\bar{Z}_0), j \in \mathcal{K} \right\} = \left\{ v \in \mathbb{S}^{3N-1} \ : \ v_{3j} = 0, j \in \mathcal{K} \right\}$ and the constants $\bar{\alpha}_\mathcal{K} > 0$ and $\bar{q}_\mathcal{K} > 0$ via

$$\bar{\alpha}_\mathcal{K} := \min \left\{ \bar{v}^\top \bar{\mathrm{H}}(T_0, \bar{Z}_0) \bar{v} \ : \ \bar{v} \in \bar{\mathbb{S}}_\mathcal{K}^{3N-1} \right\}, \tag{6.25}$$

$$\bar{q}_\mathcal{K} := \max \left\{ \sum_{j \in \mathcal{K}} |\mathrm{D}\bar{\sigma}^{(j)}(T_0, \bar{Z}_0)\bar{v}| \|\bar{\mathrm{M}}^{(j)}(\bar{Z}_0^{(j)})\| \right. \tag{6.26}$$
$$\left. + \bar{\sigma}^{(j)}(T_0, Z_0)_+ \|\mathrm{D}\bar{\mathrm{M}}^{(j)}(\bar{Z}_0^{(j)})[\bar{v}^{(j)}, \bar{u}^{(j)}]\| \ : \ \bar{u}, \bar{v} \in \bar{\mathbb{S}}_\mathcal{K}^{3N-1} \right\}.$$

For all index sets $\mathcal{I}_c \subset \mathcal{K} \subset \mathcal{I}_c \cup \mathcal{I}_s$ we further assume

$$\bar{\alpha}_\mathcal{K} > \bar{q}_\mathcal{K}. \tag{N4}$$

For the following Lemmas 6.14–6.16 we assume the Assumptions (C0)–(C3) to hold.

Lemma 6.14 (Retranslating the test sets) *Let* $\Phi \in C^2(\mathbb{R}^{3N}, \mathbb{R}^{3N})$ *be a flattening transformation composed of single transformations* Φ_j *as we constructed them in Proposition 6.12 and 6.13. Further we assume the index set* $\mathcal{K} \subset \mathcal{I}$ *to satisfy* $\mathcal{K} \subset \mathcal{I}_c \cup \mathcal{I}_s$. *Then for* $\bar{\mathbb{S}}_{\mathcal{K}}^{3N-1} := \left\{ \bar{v} \in \mathbb{S}^{3N-1} : \bar{v}^j \in T_{\partial\bar{S}}(\bar{Z}_0^{(j)}), j \in \mathcal{K} \right\}$ *and* $v := \mathrm{D}\Phi(\bar{Z}_0)\bar{v}$ *we find the relation*

$$\bar{v} \in \bar{\mathbb{S}}_{\mathcal{K}}^{3N-1} \Leftrightarrow v \in \mathbb{S}_{\mathcal{K}}^{3N-1}.$$

Proof: First by our constructions of the single transformations Φ_j in Proposition 6.12 and 6.13 we find $\mathrm{D}\Phi_j \in \mathbb{R}^{3\times3}$ to be orthonormal for all $j \in \mathcal{I}$ and thus

$$\bar{v} \in \bar{\mathbb{S}}^{3N-1} \Leftrightarrow v \in \mathbb{S}^{3N-1}.$$

The single transformations Φ_j satisfy by their Definition 6.7 the equality $\mathrm{D}\Phi_j(\bar{z}^{(j)})T_{\partial\bar{S}}(\bar{z}^{(j)}) = T_{\partial S}(\Phi_j(\bar{z}^{(j)}))$ for $\bar{z}^{(j)} \in \partial\bar{S}$ or equivalently $Z_0^{(j)} = \Phi_j(\bar{Z}_0^{(j)}) \in \partial S$. This shows

$$\bar{v} \in T_{\partial\bar{S}}(\bar{Z}_0^{(j)}) \Leftrightarrow v \in T_{\partial S}(Z_0^{(j)})$$

and thus completes the proof. ∎

Lemma 6.15 (Retranslating the energetic term) *Let* $\Phi \in C^2(\mathbb{R}^{3N}, \mathbb{R}^{3N})$ *be a flattening transformations composed of single transformations* Φ_j *as we constructed them in Proposition 6.12 and 6.13. Further we assume the index set* $\mathcal{K} \subset \mathcal{I}$ *to satisfy* $\mathcal{K} \subset \mathcal{I}_c \cup \mathcal{I}_s$. *Then for* $\bar{v} \in \bar{\mathbb{S}}_{\mathcal{K}}^{3N-1}$ *and* $v := \mathrm{D}\Phi(\bar{Z}_0)\bar{v} \in \mathbb{S}_{\mathcal{K}}^{3N-1}$ *we find*

$$\bar{v}^\top \bar{\mathrm{H}}(T_0, \bar{Z}_0)\bar{v} = v^\top \mathrm{H}(T_0, Z_0)v + \sum_{j \in \mathcal{K}} \sigma^{(j)}(T_0, Z_0) \left(v^{(j)}\right)^\top B_j v^{(j)}$$

with the matrices $B_j \in \mathbb{R}^{3\times3}$ *as defined in* (6.5).

Proof: We recall the definition of $\bar{\mathcal{E}}(T_0, \cdot) := \mathcal{E}(T_0, \Phi(\cdot))$. Hence, applying twice the chain rule we deduce the equation

$$\bar{\mathrm{H}}(T_0, \bar{Z}_0) = (\mathrm{D}\Phi(\bar{Z}_0))^\top \mathrm{H}(T_0, Z_0)\mathrm{D}\Phi(\bar{Z}_0) + \mathrm{D}\mathcal{E}(T_0, Z_0)\mathrm{D}^2\Phi(\bar{Z}_0).$$

The first term on the right hand side is already as desired. The second term is a bilinear mapping which maps $\mathbb{R}^{3N} \times \mathbb{R}^{3N}$ on $\mathbb{R}$. More in detail $\mathrm{D}^2\Phi(\bar{Z}_0)$ maps $\mathbb{R}^{3N} \times \mathbb{R}^{3N}$ on $\mathbb{R}^{3N}$ and afterwards the force $\mathrm{D}\mathcal{E}(T_0, Z_0) \in \mathbb{R}^{3N^*}$ maps $\mathbb{R}^{3N}$ on $\mathbb{R}$. For $\bar{v} \in \mathbb{R}^{3N}$ we find due to the special structure of Φ the formula $(\mathrm{D}^2\Phi)^{(j)}(\bar{Z}_0)[\bar{v}, \bar{v}] = \mathrm{D}^2\Phi_j(\bar{Z}_0^{(j)})[\bar{v}^{(j)}, \bar{v}^{(j)}] \in \mathbb{R}^3$ for all $j = 1, \ldots, N$. Hence for general $\bar{v} \in \mathbb{R}^{3N}$ we have the formula

$$\mathrm{D}\mathcal{E}(T_0, Z_0)\mathrm{D}^2\Phi(\bar{Z}_0)[\bar{v}, \bar{v}] = \sum_{j \in \mathcal{I}} (\mathrm{D}\mathcal{E})^{(j)}(T_0, Z_0)\mathrm{D}^2\Phi_j(\bar{Z}_0^{(j)})[\bar{v}^{(j)}, \bar{v}^{(j)}]. \qquad (6.27)$$

We now make use of our initial Assumption (C3) which implies $(D\mathcal{E})^{(j)}(T_0, Z_0) = 0 \in \mathbb{R}^{3^*}$ for all $j \in \mathcal{I}_s \cup \mathcal{I}_f$. This simplifies the above formula and we can replace $\sum_{j \in \mathcal{I}}$ by $\sum_{j \in \mathcal{K}}$, since $\mathcal{I}_c \subset \mathcal{K}$. Because of the assumptions $\mathcal{K} \subset \mathcal{I}_c \cup \mathcal{I}_s$ and $\bar{v} \in \bar{\mathbb{S}}_{\mathcal{K}}^{3N-1}$ we have $(\bar{v}^{(j)})_3 = 0$ for $j \in \mathcal{K}$. Since for $j \in \mathcal{I}_c \cup \mathcal{I}_s$ we constructed the single transformation Φ_j as in Proposition 6.13 this allows us to apply the formula (6.22) which we recall here

$$D^2\Phi_j(\bar{Z}_0^{(j)})[\bar{v}^{(j)}, \bar{v}^{(j)}] = -\left((\bar{v}^{(j)})^\top \begin{pmatrix} \partial_{z_1^2}^2\varphi_j(0) & \partial_{z_1 z_2}^2\varphi_j(0) & 0 \\ \partial_{z_2 z_1}^2\varphi_j(0) & \partial_{z_2^2}^2\varphi_j(0) & 0 \\ 0 & 0 & 0 \end{pmatrix} \bar{v}^{(j)}\right)\nu(Z_0^{(j)}).$$

Further we have shown in the same proposition that $D\Phi_j(\bar{Z}_0^{(j)}) = A_j \in \mathbb{R}^{3\times3}$ with A_j being the orthonormal rotation that corresponds to $Z_0^{(j)} \in \partial\mathcal{S}$. Thus replacing $\bar{v}^{(j)}$ by $\bar{v}^{(j)} = A_j^\top v^{(j)}$ provides with with the desired formula

$$D\mathcal{E}(T_0, Z_0)D^2\Phi(\bar{Z}_0)[\bar{v}, \bar{v}] = \sum_{j \in \mathcal{I}} -(v^{(j)})^\top B_j v^{(j)} \underbrace{(D\mathcal{E})^{(j)}(T_0, Z_0)\nu(Z_0^{(j)})}_{=-\sigma^{(j)}(T_0, Z_0)}$$

with $B_j = A_j \begin{pmatrix} \partial_{z_1^2}^2\varphi_j(0) & \partial_{z_1 z_2}^2\varphi_j(0) & 0 \\ \partial_{z_2 z_1}^2\varphi_j(0) & \partial_{z_2^2}^2\varphi_j(0) & 0 \\ 0 & 0 & 0 \end{pmatrix} A_j^\top \in \mathbb{R}^{3\times3}$ as defined in (6.5). ∎

Before we translate the frictional terms and show that they formally remain unchanged we recall their definitions. We are interested in the following two terms $\left|D\sigma^{(j)}(T_0, Z_0)u\right| \|M^{(j)}(Z_0^{(j)})v^{(j)}\|$ and $\sigma^{(j)}(T_0, Z_0)_+ \|DM^{(j)}(Z_0^{(j)})[v^{(j)}, u^{(j)}]\|$. as they appear in the definition of $q_\mathcal{K}$, see Assumption (6.7).

Lemma 6.16 (Retranslating the frictional terms) *Let $\Phi \in C^2(\mathbb{R}^{3N}, \mathbb{R}^{3N})$ be a flattening transformations composed of single transformations Φ_j as we constructed them in Proposition 6.12 and 6.13. Further we assume the index set $\mathcal{K} \subset \mathcal{I}$ to satisfy $\mathcal{K} \subset \mathcal{I}_c \cup \mathcal{I}_s$. Then for $\bar{u}, \bar{v} \in \bar{\mathbb{S}}_{\mathcal{K}}^{3N-1}$ and $u := D\Phi(\bar{Z}_0)\bar{u}, v := D\Phi(\bar{Z}_0)\bar{v} \in \mathbb{S}_{\mathcal{K}}^{3N-1}$ and $j \in \mathcal{K}$ we formally find*

$$D\bar{\sigma}^{(j)}(T_0, \bar{Z}_0)\bar{u}\|\bar{M}^{(j)}(\bar{Z}_0^{(j)})\bar{v}^{(j)}\| = D\sigma^{(j)}(T_0, Z_0)u\|M^{(j)}(Z_0^{(j)})v^{(j)}\| \quad (6.28)$$

$$\bar{\sigma}^{(j)}(T_0, \bar{Z}_0)\left\|D\bar{M}^{(j)}(\bar{Z}_0^{(j)})[\bar{v}^{(j)}, \bar{u}^{(j)}]\right\| = \sigma^{(j)}(T_0, Z_0)\left\|DM^{(j)}(Z_0^{(j)})[v^{(j)}, u^{(j)}]\right\|. \quad (6.29)$$

In the case of a single particle (i.e. $\mathcal{I} = \{1\}$) we present the result more in detail, i.e.

$$\bar{\sigma}(T_0, \bar{Z}_0)\frac{(\bar{M}(\bar{Z}_0)\bar{v})^\top}{\|\bar{M}(\bar{Z}_0)\bar{v}\|}D\bar{M}(\bar{Z}_0)[\bar{v}, \bar{u}] = \sigma(T_0, Z_0)\frac{(M(Z_0)v)^\top}{\|M(Z_0)v\|}DM(Z_0)[v, u]. \quad (6.30)$$

Proof: Before we start the proof we want to mention that equation (6.30) holds also in the general N-particle situation, but since we need this relation later on only for $N = 1$ this allows us to drop all indices j in the presentation of the formula. We start with the normal forces $\sigma^{(j)}$ and $\bar{\sigma}^{(j)}$. Due to $j \in \mathcal{K}$ and $\mathcal{K} \subset \mathcal{I}_c \cup \mathcal{I}_s$ we know that the j-th particle $Z_0^{(j)}$ is in contact with the boundary, i.e. $Z_0^{(j)} \in \partial \mathcal{S}$. The same holds for the corresponding 'flat' particle, i.e. $\bar{Z}_0^{(j)} \in \partial \bar{\mathcal{S}}$. By definition we have for $\bar{z}^{(j)} \in \partial \bar{\mathcal{S}}$ the following relation between the normal forces

$$
\begin{aligned}
\bar{\sigma}^{(j)}(T_0, \bar{z}) &= \left(\mathrm{D}\bar{\mathcal{E}}\right)^{(j)}(T_0, \bar{z}) \begin{pmatrix} 0 & 0 & -1 \end{pmatrix}^\top \\
&= (\mathrm{D}\mathcal{E}(T_0, \Phi(\bar{z}))\mathrm{D}\Phi(\bar{z}))^{(j)} \begin{pmatrix} 0 & 0 & -1 \end{pmatrix}^\top \\
&= (\mathrm{D}\mathcal{E})^{(j)}(T_0, \Phi(\bar{z}))\mathrm{D}\Phi_j(\bar{z}^{(j)}) \begin{pmatrix} 0 & 0 & -1 \end{pmatrix}^\top = \sigma^{(j)}(T_0, \Phi(\bar{z})).
\end{aligned}
$$

For the third and forth equality we used that the flattening transformation Φ is composed by single transformations Φ_j which satisfy $\mathrm{D}\Phi_j(\bar{z}^{(j)}) \begin{pmatrix} 0 & 0 & -1 \end{pmatrix}^\top = \nu\left(\Phi_j(\bar{z}^{(j)})\right) = \nu\left(\Phi^{(j)}(\bar{z})\right)$. Due to $\Phi(\bar{Z}_0) = Z_0$ we find on the one hand the equality of the normal forces, i.e. $\bar{\sigma}^{(j)}(T_0, \bar{Z}_0) = \sigma^{(j)}(T_0, Z_0)$ which we will use for the equations (6.29) and (6.30). On the other hand we have by the chain rule and the definition $u := \mathrm{D}\Phi(\bar{Z}_0)\bar{u}$ the equality

$$
\mathrm{D}\bar{\sigma}^{(j)}(T_0, \bar{Z}_0)\bar{u} = \mathrm{D}\sigma^{(j)}(T_0, Z_0)u
$$

which we use for equation (6.28).

We now turn towards the terms involving the friction matrices $\mathrm{M}^{(j)}$. Due to the definition of the 'flat' friction matrix $\bar{\mathrm{M}}^{(j)}$ we find the equality $\bar{\mathrm{M}}(\bar{Z}_0^{(j)})\bar{v}^{(j)} = \mathrm{M}(Z_0^{(j)})v^{(j)}$ which we use for the equations (6.28) and (6.30). Hence, it is sufficient to prove

$$
\mathrm{D}\bar{\mathrm{M}}^{(j)}(\bar{Z}_0^{(j)})[\bar{v}^{(j)}, \bar{u}^{(j)}] = \mathrm{D}\mathrm{M}^{(j)}(Z_0^{(j)})[v^{(j)}, u^{(j)}] \quad \in \mathbb{R}^{3^*}.
$$

Note that due to $j \in \mathcal{K}$ we have $\bar{u}^{(j)}, \bar{v}^{(j)} \in \mathcal{T}_{\partial \bar{\mathcal{S}}}(\bar{Z}_0^{(j)})$ or equivalently $\bar{u}_3^{(j)} = \bar{v}_3^{(j)} = 0$. In the rest of the proof we drop all indices j to keep the notation easily readable. Hence, our aim is to prove

$$
\mathrm{D}\bar{\mathrm{M}}(\bar{Z}_0)[\bar{v}, \bar{u}] = \mathrm{D}\mathrm{M}(Z_0)[v, u] \quad \in \mathbb{R}^{3^*}. \tag{6.31}
$$

By definition we have $\bar{\mathrm{M}}(\bar{z}) = \mathrm{M}(\Phi(\bar{z}))\mathrm{D}\Phi(\bar{z})$ and using the product rule we calculate

$$
\mathrm{D}\bar{\mathrm{M}}(\bar{Z}_0)[\bar{v}, \bar{u}] = \mathrm{D}\mathrm{M}(Z_0)[\mathrm{D}\Phi(\bar{Z}_0)\bar{v}, \mathrm{D}\Phi(\bar{Z}_0)\bar{u}] + \mathrm{M}(Z_0)\mathrm{D}^2\Phi(\bar{Z}_0)[\bar{v}, \bar{u}] \quad \in \mathbb{R}^{3^*} \tag{6.32}
$$

for arbitrary $\bar{v}, \bar{u} \in \mathbb{R}^3$. Due to $v = \mathrm{D}\Phi(\bar{Z}_0)\bar{v}$ and $u = \mathrm{D}\Phi(\bar{Z}_0)\bar{u}$ the first term supplies us with the desired expression. Consequently we establish the result if we show that the second term $\mathrm{M}(Z_0)\mathrm{D}^2\Phi(\bar{Z}_0)[\bar{v}, \bar{u}]$ in (6.32) equals zero for test vectors

$\bar{v}, \bar{u} \in \mathbb{R}^3$ satisfying $\bar{v}_3 = \bar{u}_3 = 0$. For such vectors we can apply formula (6.22) of Proposition 6.13. (Recall that the index j which we dropped satisfied $j \in \mathcal{K}$ and due to $\mathcal{K} \subset \mathcal{I}_c \cup \mathcal{I}_s$ we see that $Z_0^{(j)} \in \partial \mathcal{S}$ holds. Thus Φ is defined as in Proposition 6.13.) Thus we find

$$\mathrm{M}(Z_0)\mathrm{D}^2\Phi(\bar{Z}_0)[\bar{v}, \bar{u}] = -\underbrace{\mathrm{M}(Z_0)\nu(Z_0)}_{=0} \sum_{ij=1}^{2} \partial_{z_{ij}}^2 \varphi_0(0)\bar{v}_i\bar{u}_j = 0$$

since the matrix of friction M satisfies $\mathrm{M}(Z_0)\nu(Z_0) = 0 \in \mathbb{R}^3$, see Assumption (C2).
∎

Corollary 6.17 (Proof of Theorem 6.5) *Let the Assumptions* (C0)–(C4) *hold then there exists a solution of the curved Problem 6.4.*

Proof: The last three Lemmas 6.14–6.16 together show

$$\alpha_{\mathcal{K}} = \bar{\alpha}_{\mathcal{K}} \quad \text{and} \quad \mathsf{q}_{\mathcal{K}} = \bar{\mathsf{q}}_{\mathcal{K}}.$$

The constants $\alpha_{\mathcal{K}}$ and $\bar{\alpha}_{\mathcal{K}}$ denote the lower bound for the change of the elastic forces in either the flat or curved case and are defined in (6.6) and (6.25). The constants $\mathsf{q}_{\mathcal{K}}$ and $\bar{\mathsf{q}}_{\mathcal{K}}$ defined in (6.7) and (6.26)) describe the upper bound for the change of the frictional forces. This finally proves that the curved and flat convexity Assumptions (C4) and (N4), i.e. $\alpha_{\mathcal{K}} > \mathsf{q}_{\mathcal{K}}$ and $\bar{\alpha}_{\mathcal{K}} > \bar{\mathsf{q}}_{\mathcal{K}}$, are equivalent. This was the last missing step since we have shown in Corollary 6.11 that the curved Assumptions (C0)–(C3) imply the flat Assumptions (N0)–(N3). Thus, under the Assumptions (C0)–(C4) we can apply Theorem 5.2 and establish the existence of a solution of the flat Problem 6.9 which is equivalent to the existence of a solution of the curved Problem 6.4, see Theorem 6.10. This proves the existence Theorem 6.5.
∎

6.6   Special case $N = 1$

In Chapter 4 we established existence of a solution in the special case of a single particle and a flat boundary. There we used a weaker convexity assumption than for the corresponding N particle case. We are now interested how this weak convexity assumption reads in the curved situation. Since the one particle case is contained in the N particle case we proceed as in the previous Sections 6.2–6.5 to construct a solution. We will content ourself with translating the flat assumptions only.

We quickly present the 'curved' assumptions which are mainly the same as in the N particle case apart from the assumption on the friction matrix M and the convexity assumption. For the admissible set $\mathcal{A} \subset \mathbb{R}^3$ with $\mathcal{A} \neq \mathbb{R}^3$ and we assume

$$\mathcal{A} = \mathrm{clos}(\mathrm{int}\mathcal{A}) \quad \text{and} \quad \partial\mathcal{A} \text{ belongs locally to } \mathrm{C}^3 \tag{CO0}$$

and for the energy functional

$$\mathcal{E} \in C^2\big([0,T] \times \mathcal{A}, [0,\infty)\big). \tag{CO1}$$

As usual in a one particle situation we exclude for technical simplicity of the proof the possibility of having $M(z) = 0 \in \mathbb{R}^{3\times 3}$ for some $z \in \partial\mathcal{A}$ and thus claim

$$M \in C^1\big(\partial\mathcal{A}, \mathbb{R}^{3\times 3}\big) \text{ with } M(z)\nu(z) = 0 \text{ for all } z \in \partial\mathcal{A} \tag{CO2}$$
$$\text{and } \min\big\{\|M(Z_0)v\| \ : \ v \in \mathcal{T}_{\partial\mathcal{A}}(Z_0) \cap \mathbb{S}^2\big\} > 0.$$

The dissipation functional Ψ and set of normal forces $\mathcal{N}_\mathcal{A}$ are defined as in the curved N particle case taking $N = 1$. We have the classical problem formulation.

Problem 6.18 *For given initial time $T_0 \in [0,T)$ and initial state $Z_0 \in \mathcal{A}$ find a positive time span $\Delta \in (0, T-T_0]$ and a solution $z \in W^{1,\infty}([T_0, T_0+\Delta], \mathcal{A})$ such that the initial condition $z(T_0) = Z_0$ is satisfied and such that for almost all $t \in [T_0, T_0+\Delta]$ the following differential inclusion holds*

$$0 \in D\mathcal{E}(t, z(t)) + \partial_v\Psi(t, z(t), \dot{z}(t)) + \mathcal{N}_\mathcal{A}(z(t)) \quad \subset \mathbb{R}^{3^*}. \tag{DI}$$

The usual initial Assumption reads

$$0 \in D\mathcal{E}(T_0, Z_0) + \partial_v\Psi(T_0, Z_0, 0) + \mathcal{N}_\mathcal{A}(Z_0) \subset \mathbb{R}^{3^*}. \tag{CO3}$$

The convexity assumption in the N particle case includes a case study by different choices of the index set $\mathcal{K}$. As usual we treat the convexity assumption in the one particle case in detail and we distinguish between different assumptions. Hence, for the initial situation with no normal force we assume that there exists a constant $\alpha_* > 0$ such that

$$v^\top H(T_0, Z_0)v \geq \alpha_*\|v\|^2 \quad \text{for all } v \in \mathbb{R}^3. \tag{CO4}$$

If the particle is in contact we have to replace the Hessian matrix H by $H(T_0, Z_0) + \sigma(T_0, Z_0)B_0$ with the matrix

$$B_0 := A_0 \begin{pmatrix} \partial^2_{z_1^2}\varphi_0(0) & \partial^2_{z_1 z_2}\varphi_0(0) & 0 \\ \partial^2_{z_2 z_1}\varphi_0(0) & \partial^2_{z_2^2}\varphi_0(0) & 0 \\ 0 & 0 & 0 \end{pmatrix} A_0^\top.$$

Here A_0 and φ_0 are the rotation and shape function that correspond to the initial value $Z_0 \in \partial\mathcal{A}$, see also (6.5). Hence in the initial situation of contact we make the joint convexity assumption

$$v^\top\Big(H(T_0, Z_0) + \sigma(T_0, Z_0)B_0\Big)v$$
$$+\sigma(T_0, Z_0)DM(Z_0)\frac{(M(Z_0)v)^\top}{\|M(Z_0)v\|}[v, v] + \|M(Z_0)v\|D\sigma(T_0, Z_0)v > 0 \quad \text{(CO5)}$$

for all $v \in \mathcal{T}_{\partial\mathcal{A}}(Z_0) \cap \mathbb{S}^2$. Here we recall the notational convention $\big(DM(z)[u,v]\big)_i = \sum_{j,k=1}^3 \partial_{z_k}M_{ij}(z)u_j v_k$ for $i = 1, \ldots, 3$ and $u, v \in \mathbb{R}^3$.

Theorem 6.19 (Existence) *Let the Assumptions* (CO0)–(CO3) *hold.*

1. **(no contact)** *If we have $Z_0 \in \text{int} \mathcal{A}$ we assume the convexity Assumption* (CO4) *on $\mathcal{E}$,*

2. **(positive normal force)** *if we have $-\mathrm{D}\mathcal{E}(T_0, Z_0)\nu(Z_0) > 0$ (this implies $Z_0 \in \partial\mathcal{A}$ due to* (CO3)*) we assume the joint convexity* (CO5) *of $\mathcal{E}$ and Ψ,*

3. **(contact and no normal force)** *if we have $Z_0 \in \partial\mathcal{A}$ and $\mathrm{D}\mathcal{E}(T_0, Z_0)\nu(Z_0) = 0$ we assume* (CO4) *and* (CO5),

then there exists a positive time span $\Delta > 0$ and a solution $z \in \mathrm{W}^{1,\infty}([T_0, T_0+\Delta], \mathcal{A})$ of the 'curved' Problem 6.18.

Proof: The curved Problem 6.18 is equivalent to the flat single particle Problem 4.1, see Theorem 6.10. Hence, by the single particle existence Theorem 4.2 we derive existence of a solution if we assure the flat Assumptions (O0)–(O5) for the flat data to hold. Taking $N = 1$ in Corollary 6.11 we directly see that the curved Assumptions (CO0)–(CO3) imply the flat Assumptions (O0)–(O3). The only detail we have to check is that the inequality

$$\min\left\{\|\bar{\mathrm{M}}(\bar{Z}_0)\bar{v}\| \,:\, \bar{v} \in \mathcal{T}_{\partial\bar{\mathcal{A}}}(\bar{Z}_0) \cap \mathbb{S}^2\right\} > 0 \tag{6.33}$$

holds. This inequality is part of the flat Assumption (O2) but not of (N2) if we choose $N = 1$. We just recall the facts that the flattening transformation satisfies $\mathrm{D}\Phi(\bar{Z}_0)\left(\mathcal{T}_{\partial\bar{\mathcal{A}}}(\bar{Z}_0) \cap \mathbb{S}^2\right) = \mathcal{T}_{\partial\mathcal{A}}(Z_0) \cap \mathbb{S}^2$, that the flat matrix $\bar{\mathrm{M}}$ is defined by $\bar{\mathrm{M}}(\bar{Z}_0) = \mathrm{M}(Z_0)\mathrm{D}\Phi(\bar{Z}_0)$ and that the Assumption (CO2) includes the inequality $\min\{\|\mathrm{M}(Z_0)v\| \,:\, v \in \mathcal{T}_{\partial\mathcal{A}}(Z_0) \cap \mathbb{S}^2\} > 0$. Summarizing we derive inequality (6.33). Hence, we miss only the convexity Assumptions (O4) and (O5). Taking $\mathcal{K} = \emptyset$ in Lemma 6.15 we find

$$\bar{v}^\top \bar{\mathrm{H}}(T_0, \bar{Z}_0)\bar{v} = v^\top \mathrm{H}(T_0, Z_0)v$$

for all $\bar{v} \in \mathbb{S}^2$ and $v := \mathrm{D}\Phi(\bar{Z}_0)\bar{v} \in \mathbb{S}^2$. This shows that the convexity Assumptions (CO4) and (O4) coincide. The joint convexity Assumption (O5) is only assumed in the flat existence Theorem 4.2 in the case of having contact at time $t = T_0$, i.e. either $\mathcal{I}_c = \{1\}$ or $\mathcal{I}_s = \{1\}$ holds. We thus can take $\mathcal{K} = \{1\}$ in Lemma 6.15 and find for the elastic terms

$$\bar{v}^\top \bar{\mathrm{H}}(T_0, \bar{Z}_0)\bar{v} = v^\top \left(\mathrm{H}(T_0, Z_0) + \sigma(T_0, Z_0)B_0\right)v$$

for $\bar{v} \in \mathcal{T}_{\partial\bar{\mathcal{A}}}(\bar{Z}_0) \cap \mathbb{S}^2$ and $v := \mathrm{D}\Phi(\bar{Z}_0)\bar{v} \in \mathcal{T}_{\partial\mathcal{A}}(Z_0) \cap \mathbb{S}^2$. For the equivalence of the frictional terms we use Lemma 6.16 and more precisely equation (6.30) there. Taking the same test vectors as above we deduce

$$\bar{\sigma}(T_0, \bar{Z}_0)\frac{(\bar{\mathrm{M}}(\bar{Z}_0)\bar{v})^\top}{\|\bar{\mathrm{M}}(\bar{Z}_0)\bar{v}\|}\mathrm{D}\bar{\mathrm{M}}(\bar{Z}_0)[\bar{v}, \bar{v}] = \sigma(T_0, Z_0)\frac{(\mathrm{M}(Z_0)v)^\top}{\|\mathrm{M}(Z_0)v\|}\mathrm{D}\mathrm{M}(Z_0)[v, v].$$

Since $D\Phi(\bar{Z}_0)$ maps the set $\mathcal{T}_{\partial\bar{\mathcal{A}}}(\bar{Z}_0)\cap\mathbb{S}^2$ bijectively on the set $\mathcal{T}_{\partial\mathcal{A}}(Z_0)$ the last two equations establish the equivalence of the Assumptions (CO5) and (O5). We ended retranslating the Assumptions and summarizing find that the curved Assumptions (CO0)–(CO5) imply the flat Assumptions (O0)–(O5). With this Assumptions we finally establish the existence of a solution, see Theorems 4.2 and 6.10. ∎

Appendix A

Subdifferential calculus

In the Appendix we present some results from convex analysis which are reused in the whole thesis. For the convenience of the reader we repeat the Definition 2.1 of a subdifferential.

Definition A.1 (Subdifferential) *Let* $f : \mathbb{R}^n \to (-\infty, \infty]$ *for* $z \in \mathbb{R}^n$ *we define*

$$\partial f(z) := \left\{ v^* \in \mathbb{R}^{n^*} : v^*(y - z) \leq f(y) - f(z) \text{ for all } y \in \mathbb{R}^n \right\}.$$

We call the set $\partial f(z) \subset \mathbb{R}^{n^*}$ *the subdifferential of* f *in* z.

We now present two classical results from convex analysis. Both can be found in the book [EkT76], see Proposition 5.3 and 5.6 there.

Theorem A.2 *Let* $f : \mathbb{R}^n \to \mathbb{R}$ *be convex and assume for* $z \in \mathbb{R}^n$ *the Gateaux-derivative* $\mathrm{D}f(z) \in \mathbb{R}^{n^*}$ *to exist then we have*

$$\partial f(z) = \{\mathrm{D}f(z)\}.$$

We present the following result in a simplified form. It was established by Moreau [Mor66] and Rockafellar [Roc66].

Theorem A.3 (Moreau-Rockafellar 1966) *For* $j = 1, 2$ *let* $f_j : \mathbb{R}^n \to (-\infty, \infty]$ *be a convex, lower semi-continuous function. If* $\max\{f_1(z), f_2(z)\} < \infty$ *holds for some point* $z \in \mathbb{R}^n$ *and* f_1 *is continuous in* z, *then we have:*

$$\partial(f_1 + f_2)(z) = \partial f_1(z) + \partial f_2(z).$$

In the following lemma we prove a summation formula for subdifferentials similar to the formula found in the *Moreau-Rockafellar*-theorem. This time we prove the summation formula for a concrete example for which the assumptions of the *Moreau-Rockafellar*-theorem does not hold since one can choose the argument z such that all functions are not continuous in z.

Lemma A.4 *For $a, b \in [-\infty, \infty]^n$ with $a_j \leq b_j$ for all $j = 1, \ldots, n$ we define the stripes $\mathcal{S}_j := \{z \in \mathbb{R}^n : a_j \leq z_j \leq b_j\}$ and the cuboid $\mathcal{Q} = \cap_{j=1}^n \mathcal{S}_j$. Then for $z \in \mathcal{Q}$ we find the formula*

$$\partial \mathcal{X}_{\mathcal{Q}}(z) = \sum_{j=1}^n \partial \mathcal{X}_{\mathcal{S}_j}(z). \tag{A.1}$$

Proof: Since we have $\mathcal{X}_{\mathcal{Q}} = \sum_{j=1}^n \mathcal{X}_{\mathcal{S}_j}$ formula (A.1) is equivalent to the equality $\partial(\sum_{j=1}^n \mathcal{X}_{\mathcal{S}_j})(z) = \sum_{j=1}^n \partial \mathcal{X}_{\mathcal{S}_j}(z)$. Let a general index set $\mathcal{J} \subset \{1, \ldots, n\}$ and $j_* \in \mathcal{J}$ be given. We will show

$$\partial \left(\sum_{j \in \mathcal{J}} \mathcal{X}_{\mathcal{S}_j} \right)(z) = \partial \left(\sum_{j \in \mathcal{J}/\{j_*\}} \mathcal{X}_{\mathcal{S}_j} \right)(z) + \partial \mathcal{X}_{\mathcal{S}_{j_*}}(z) \tag{A.2}$$

since applying this formula recursively establishes (A.1). The inclusion '$\supset$' holds for general convex functions and can be directly checked. The aim is to establish the inverse inclusion '$\subset$' which in general does not hold. Here we will have to exploit the special structure of our functions. Let $v^* \in \partial \left(\sum_{j \in \mathcal{J}} \mathcal{X}_{\mathcal{S}_j} \right)(z) \subset \mathbb{R}^{n^*}$. We split this vector, i.e. we write $v^* = v_{\Sigma}^* + v_*^*$ with the vectors $v_{\Sigma}^*, v_*^* \in \mathbb{R}^{n^*}$ being defined via

$$(v_{\Sigma}^*)_j := \begin{cases} (v^*)_j & \text{if } j \neq j_*, \\ 0 & \text{if } j = j_* \end{cases}$$

and $v_*^* := v^* - v_{\Sigma}^*$. We prove the inclusion '$\subset$' if we show $v_{\Sigma}^* \in \partial \left(\sum_{j \in \mathcal{J}/\{j_*\}} \mathcal{X}_{\mathcal{S}_j} \right)(z)$ and $v_*^* \in \partial \mathcal{X}_{\mathcal{S}_{j_*}}(z)$. By definition the first inclusion is equivalent to the inequality $v_{\Sigma}^* v \leq \sum_{j \in \mathcal{J}/\{j_*\}} \mathcal{X}_{\mathcal{S}_j}(z + v)$ for all $v \in \mathbb{R}^n$. We fix $v \in \mathbb{R}^n$ and define $v_{\Sigma} \in \mathbb{R}^n$ via $(v_{\Sigma})_j := v_j$ for $j \neq j_*$ and $v_{j_*} := 0$. We then find

$$v_{\Sigma}^* v = v^* v_{\Sigma} \leq \sum_{j \in \mathcal{J}} \mathcal{X}_{\mathcal{S}_j}(z + v_{\Sigma}) = \sum_{j \in \mathcal{J} \backslash j_*} \mathcal{X}_{\mathcal{S}_j}(z + v_{\Sigma}) = \sum_{j \in \mathcal{J} \backslash j_*} \mathcal{X}_{\mathcal{S}_j}(z + v).$$

The above inequality is due to the definition of v^*. The last two equalities follow from the structure of the stripes $\mathcal{S}_j$ as follows. Due to $(v_{\Sigma})_{j_*} = 0$ we have that $0 = \mathcal{X}_{\mathcal{S}_{j_*}}(z)$ implies $0 = \mathcal{X}_{\mathcal{S}_{j_*}}(z + v_{\Sigma})$. In the last equality we exploited $(z + v_{\Sigma})_j = (z + v)_j$ for $j \neq j_*$. The proof of $v_*^* \in \partial \mathcal{X}_{c \mathcal{S}_{j_*}}(z)$ is analog. We thus established the equation (A.1). $\blacksquare$

Bibliography

[AK*02] A. AMASSAD, K. L. KUTTLER, M. ROCHDI, and M. SHILLOR. Quasi-static thermoviscoelastic contact problem with slip dependent friction co-efficient. *Math. Comput. Modelling*, 36(7-8), 839–854, 2002. Lyapunov's methods in stability and control.

[AK*05] L.-E. ANDERSSON, A. KLARBRING, J. R. BARBER, and M. CIAVARELLA. On the existence and uniqueness of steady state solutions in thermoelastic contact with frictional heating. *Proc. R. Soc. Lond. Ser. A Math. Phys. Eng. Sci.*, 461(2057), 1261–1282, 2005.

[AMS08] F. AURICCHIO, A. MIELKE, and U. STEFANELLI. A rate-independent model for the isothermal quasi-static evolution of shape-memory materials. *Math. Models Methods Appl. Sci.*, 18(1), 125–164, 2008.

[And99] L.-E. ANDERSSON. Quasistatic fricional contact problem with finitely many degrees of freedom. LiTH-MAT-R-1999-22,Department of Mathematics, 1999.

[And00] L.-E. ANDERSSON. Existence results for quasistatic contact problems with coulomb friction. *Appl. Math. Optim.*, 42(2), 169–202, 2000.

[AnR06] L.-E. ANDERSSON and A. RIETZ. Noncoercive incremental friction problems for discrete systems. *Zeitschrift für angewandte Mathematik und Mechanik*, pages 1–19, 2006.

[Bal99] P. BALLARD. A counter-example to uniqueness in quasi-static elastic contact problems with small friction. *Internat.J.Engrg.Sci.*, 37(2), 163–178, 1999.

[BKS04] M. BROKATE, P. KREJČÍ, and H. SCHNABEL. On uniqueness in evolution quasivariational inequalities. *J. Convex Anal.*, 11(1), 111–130, 2004.

[ChA04] O. CHAU and B. AWBI. Quasistatic thermoviscoelastic frictional contact problem with damped response. *Appl. Anal.*, 83(6), 635–648, 2004.

[CHM02] C. CARSTENSEN, K. HACKL, and A. MIELKE. Non-convex potentials and microstructures in finite-strain plasticity. *R. Soc. Lond. Proc. Ser. A Math. Phys. Eng. Sci.*, 458(2018), 299–317, 2002.

[Dac08] B. DACOROGNA. *Direct methods in the calculus of variations*, volume 78 of *Applied Mathematical Sciences*. Springer, New York, second edition, 2008.

[DaZ07] G. DAL MASO and C. ZANINI. Quasi-static crack growth for a cohesive zone model with prescribed crack path. *Proc. Roy. Soc. Edinburgh Sect. A*, 137(2), 253–279, 2007.

[DDM06] G. DAL MASO, A. DESIMONE, and M. G. MORA. Quasistatic evolution problems for linearly elastic-perfectly plastic materials. *Arch. Ration. Mech. Anal.*, 180(2), 237–291, 2006.

[DFT05] G. DAL MASO, G. A. FRANCFORT, and R. TOADER. Quasistatic crack growth in nonlinear elasticity. *Arch. Ration. Mech. Anal.*, 176(2), 165–225, 2005.

[EcJ01] C. ECK and J. JARUŠEK. On the thermal aspect of dynamic contact problems. In *Proceedings of Partial Differential Equations and Applications (Olomouc, 1999)*, volume 126, pages 337–352, 2001.

[EkT76] I. EKELAND and R. TEMAM. *Convex analysis and variational problems*. North-Holland Publishing Co., Amsterdam, 1976. Translated from the French, Studies in Mathematics and its Applications, Vol. 1.

[HSS01] W. HAN, M. SHILLOR, and M. SOFONEA. Variational and numerical analysis of a quasistatic viscoelastic problem with normal compliance, friction and damage. *J. Comput. Appl. Math.*, 137(2), 377–398, 2001.

[Kla90] A. KLARBRING. Example of nonuniqueness and nonexistence of solutions to quasi-static contact problems with friction. *Ingenieur Archiv*, 60(8), 529–541, 1990.

[KMR06] M. KOČVARA, A. MIELKE, and T. ROUBÍČEK. A rate-independent approach to the delamination problem. *Math. Mech. Solids*, 11(4), 423–447, 2006.

[KMS88] A. KLARBRING, A. MIKELIĆ, and M. SHILLOR. Frictional contact problems with normal compliance. *Internat. J. Engrg. Sci.*, 26(8), 811–832, 1988.

[KMS89] A. KLARBRING, A. MIKELIĆ, and M. SHILLOR. On friction problems with normal compliance. *Nonlinear Anal.*, 13(8), 935–955, 1989.

[KMS91] A. KLARBRING, A. MIKELIĆ, and M. SHILLOR. A global existence result for the quasistatic frictional contact problem with normal compliance. In

Unilateral problems in structural analysis, VI (Capri, 1989), pages 85–111. Internat. Ser. Numer. Math. 101. Birkhäuser, Basel, 1991.

[KMZ08] D. KNEES, A. MIELKE, and C. ZANINI. On the viscid limit of a model of crack propagation. *Math. Models Methods Appl. Sci.*, 2008. to appear.

[KnM08] D. KNEES and A. MIELKE. Energy release rate for cracks in finite-strain elasticity. *Math. Models Methods Appl. Sci.*, 13(5), 501–528, 2008.

[Kre99] P. KREJČÍ. Evolution variational inequalities and multidimensional hysteriesis operators. In *Nonlinear Differential Equations*, page Chapter 2. Chapman & Hall/CRC, 1999.

[MaM08] A. MAINIK and A. MIELKE. Global existence for rate-independent gradient plasticity at finite strain. *J. Nonlinear Sci.*, 2008. submitted.

[Mie05] A. MIELKE. Evolution in rate-independent systems (ch.6). In C.Dafermos and E.Feireisl, editors, *Handbook of Differential Equations, Evolutionary Equations*, volume 2, pages 461–559. Elsevier B.V., 2005.

[MiP07] A. MIELKE and A. PETROV. Thermally driven phase transformation in shape-memory alloys. *Adv. Math. Sci. Appl.*, 17(2), 667–685, 2007.

[MiR07] A. MIELKE and R. ROSSI. Existence and uniqueness results for a class of rate-independent hysteresis problems. *Math. Models Methods Appl. Sci.*, 17(1), 81–123, 2007.

[MiT99] A. MIELKE and F. THEIL. A mathematical model for rate–independent phase transformations with hysteresis. In H.-D. Alber, R. Balean, and R. Farwig, editors, *Proceedings of the Workshop on "Models of Continuum Mechanics in Analysis and Engineering"*, pages 117–129. Shaker–Verlag, 1999.

[MiT04] A. MIELKE and F. THEIL. On rate-independent hysteresis models. *NoDEA Nonlinear Differential Equations Appl.*, 11(2), 151–189, 2004. Accepted July 2001.

[MMP05] J. MARTINS, M. MONTEIRO MARQUES, and A. PETROV. Dynamics with friction and persistent contact. *ZAMM Z. Angew. Math. Mech.*, 85(8), 531–538, 2005.

[MMP07] J. A. C. MARTINS, M. D. P. MONTEIRO MARQUES, and A. PETROV. On the stability of quasi-static paths for finite dimensional elastic-plastic systems with hardening. *ZAMM Z. Angew. Math. Mech.*, 87(4), 303–313, 2007.

[Mon93] M. D. P. MONTEIRO MARQUES. *Differential inclusions in nonsmooth mechanical problems :Shocks and dry friction.* Birkhäuser Verlag, Basel, 1993.

[Mor66] J. MOREAU. *Fonctionnelles convexes.* Séminaire sur les Équations aux Dérivées Partielles (1966–1967). II. Collège de France, Paris, 1966.

[MSo*05] J. MARTINS, F. SIMÕES, A. PINTO DA COSTA, and I. COELHO. Three examples on "(in)stability of quasi-static paths". Report ICIST DTC No.16/05, 2005.

[PiM03] A. PINTO DA COSTA and J. A. C. MARTINS. The evolution and rate problems and the computation of all possible evolutions in quasi-static frictional contact. *Comput. Methods Appl. Mech. Engrg.*, 192, 2791–2821, 2003.

[PMM07] A. PETROV, J. A. C. MARTINS, and M. D. P. MONTEIRO MARQUES. Mathematical results on the stability of quasi-static paths of elastic-plastic systems with hardening. In *Topics on mathematics for smart systems*, pages 167–182. World Sci. Publ., Hackensack, NJ, 2007.

[Roc66] R. T. ROCKAFELLAR. Characterization of the subdifferentials of convex functions. *Pacific J. Math.*, 17, 497–510, 1966.

[Roc81] R. T. ROCKAFELLAR. *The theory of subgradients and its applications to problems of optimization*, volume 1 of *R & E.* Heldermann Verlag, Berlin, 1981. Convex and nonconvex functions.

[Roc99] R. ROCCA. Existence of a solution for a quasistatic problem of unilateral contact with local friction. *C.R.Acad.Sci.Paris Sér. I Math.*, 328(12), 1253–1258, 1999.

[Roc01] R. ROCCA. Existence of approximation of a solution to quasi-static problem signorini problem with local friction. *Internat. J. Engrg. Sci.*, 39, 1233–1255, 2001.

[RTT06] T. RATIU, A. TIMOFTE, and V. TIMOFTE. Existence, uniqueness and regularity of solutions for a thermomechanical model of shape memory alloys. *Math. Mech. Solids*, 11(6), 563–574, 2006.

[ScM07] F. SCHMID and A. MIELKE. Existence results for a contact problem with varying friction coefficient and nonlinear forces. *ZAMM Z. Angew. Math. Mech.*, 87(8-9), 616–631, 2007.

[SMR07] F. SCHMID, J. A. C. MARTINS, and N. REBROVA. New results on the stability of quasi-static paths of a single particle system with Coulomb friction and persistent contact. In *Topics on mathematics for smart systems*, pages 208–217. World Sci. Publ., Hackensack, NJ, 2007.

[SST04] M. SHILLOR, M. SOFONEA, and J. J. TELEGA. Quasistatic viscoelastic contact with friction and wear diffusion. *Quart. Appl. Math.*, 62(2), 379–399, 2004.

[ToT07] A. TOUZALINE and D. E. TENIOU. A quasistatic unilateral contact problem with friction for nonlinear elastic materials. *Math. Model. Anal.*, 12(4), 497–514, 2007.

[Tou06a] A. TOUZALINE. A quasistatic bilateral contact problem with friction for nonlinear elastic materials. *Electron. J. Differential Equations*, pages No. 58, 9 pp. (electronic), 2006.

[Tou06b] A. TOUZALINE. A quasistatic unilateral contact problem with slip-dependent coefficient of friction for nonlinear elastic materials. *Electron. J. Differential Equations*, pages No. 144, 14 pp. (electronic), 2006.

Lebenslauf

Name: Florian Schmid
Geburtsdatum: 20.11.1975
Geburtsort: Oberndorf a. N.
Eltern: Martin Schmid (Zimmerermeister)
                 Edith Schmid geb. Kimmich (Erzieherin)

Schulausbildung:

1982 - 1986   Grundschule Fluorn-Winzeln
1986 - 1995   Gymnasium Schramberg, Abitur

Zivildienst:

1995 - 1996   Krankenhaus Oberndorf

Studium:

WS 96/97 - SS 98   Studium der Humanbiologe
                   Phillips-Universität Marburg
WS 98/99 - SS 03   Studium der Mathematik mit Nebenfach Physik und Informatik
                   Universität Stuttgart
WS 00/01 - SS 01   Studium der Mathematik und Informatik
                   INSA und Université Claude Bernard, Lyon (Frankreich)
15.5.2003   Diplom in Mathematik

Berufliche Tätigkeit:

07.2003 - 08.2005   Wissenschaftlicher Angestellter
                    Institut für Analysis, Dynamik und Modellierung
                    Universität Stuttgart
09.2005 - 02.2006   Stipendiat des EU-Programms 'Smart Systems'
                    Departamento Engenheria Civil
                    Instituto Superior Técnico, Lissabon (Portugal)
03.2006 - 06.2008   Wissenschaftlicher Angestellter
                    Arbeitsgruppe Partielle Differentialgleichungen
                    Weierstraß Institut für angewandte Analysis und Stochastik, Berlin